INSTRUCTOR'S RESOURCE GUIDE
to Accompany

Residential Construction Academy Carpentry

Second Edition

Mark Martin

with contributions by:
Floyd Vogt and Kathy Sutphin

THOMSON
DELMAR LEARNING

Australia • Brazil • Canada • Mexico • Singapore • Spain • United Kingdom • United States

THOMSON

DELMAR LEARNING

Instructor's Resource Guide to Accompany
Residential Construction Academy: Carpentry, 2e
Mark Martin, Floyd Vogt, and Kathy Sutphin

Vice President, Technology and Trades ABU:
David Garza

Director of Learning Solutions:
Sandy Clark

Managing Editor:
Larry Main

Acquisitions Editor:
Jame DeVoe

Product Manager:
John Fisher

Marketing Director:
Deborah Yarnell

Marketing Manager:
Kevin Rivenburgh

Marketing Coordinator:
Mark Pierro

Director of Production:
Patty Stephan

Production Manager:
Andrew Crouth

Content Project Manager:
Andrea Majot

Editorial Assistant:
Tom Best

Printed in the United States of America
1 2 3 4 5 XX 09 08 07

For more information contact Thomson Delmar Learning
Executive Woods
5 Maxwell Drive, PO Box 8007,
Clifton Park, NY 12065-8007

Or find us on the World Wide Web at
www.delmarlearning.com

ISBN 10: 1-4180-0184-8
ISBN 13: 978-1-4180-0184-1

NOTICE TO THE READER

Publisher does not warrant or guarantee any of the products described herein or perform any independent analysis in connection with any of the product information contained herein. Publisher does not assume, and expressly disclaims, any obligation to obtain and include information other than that provided to it by the manufacturer.

The reader is expressly warned to consider and adopt all safety precautions that might be indicated by the activities herein and to avoid all potential hazards. By following the instructions contained herein, the reader willingly assumes all risks in connection with such instructions.

The publisher makes no representation or warranties of any kind, including but not limited to, the warranties of fitness for particular purpose or merchantability, nor are any such representations implied with respect to the material set forth herein, and the publisher takes no responsibility with respect to such material. The publisher shall not be liable for any special, consequential, or exemplary damages resulting, in whole or part, from the readers' use of, or reliance upon, this material.

Table of Contents

Standards Summary

National Association of Home Builders and Home Builders Institute

Residential Carpentry Standards Overview

Working with many industry experts, the Home Builders Institute has established national standards for the residential construction industry that reflect industry skill requirements. These standards provide a basis for the certification and training of workers, and provide employers with objective benchmarks for selecting employees and evaluating training needs. In addition, educators will find the standards useful for designing curriculum and evaluating individual training outcomes.

The Residential Carpentry Standards were developed by a committee of NAHB industry leaders from various areas within the field of residential carpentry. The committee included remodelers, custom builders, small volume and production builders, carpentry instructors and trainers from various regions of the country, as well as a representative of the Vinyl Siding Institute, a carpentry trainer for a nationwide high-production home builder, and the author and editor of Delmar's *Carpentry* text.

An initial group considered performance standards, practices, and materials; reviewed and rated work functions, activities, and skills; and classified 11 carpentry specialties. A second group of industry leaders validated the work of the original team and produced a final list that includes 32 work functions and 231 key activities. In addition, the process met or exceeded the American Psychological Association standards for occupational credentialing.

- Residential Carpenter standards were developed and validated for the following 11 carpentry specialties: *Wood Framing; Metal Framing; Interior Trim; Exterior Trim; Siding; Concrete Forms; Drywall; Roof Application; Stairs; Insulation;* and *Cabinets and Countertops*. These classifications reflect the trend towards specialization in the home building industry.
- Thirty-two (32) *critical work functions,* describing the major tasks and content areas of work within each specialty, were identified. These 32 critical work functions reflect and accommodate regional and other differences throughout the industry.
- *Key activities* or major tasks and knowledge involved in completing critical work functions are also provided.
- *Applied academic* skills required to perform key activities, including measurement, arithmetic, layout, geometry, communications, and materials, are provided.
- *Safety* requirements involved in completing key activities have also been identified.

The Carpentry Skill Standards in full text can be viewed and printed from the Instructor's e.resource™ CD-ROM that accompanies the Residential Construction Academy *Carpentry* text, or at www.residentialacademy.com.

To receive a hard copy version of these standards, please contact Steven Kramer at Home Builders Institute, by phone at 202-266-8925 or by email at kramers@hbi.org.

SECTION ONE

Chapter Outlines

Chapter 1 Hand Tools

Instructor's Note

In this column, you will find a variety of lecture tips to aid student comprehension, notes on when to use PowerPoint slides to teach concepts, safety cautions for your students, and notes on when your class should view a video segment.

Appendix A lists corresponding numbered PowerPoint Presentation titles (see page 197)

Appendix B lists corresponding numbered Video titles and sections as applicable (see page 201)

Introduction

1. Construction workers need to be knowledgeable in the safe operation of a wide variety of tools.
2. It is essential to know how to select the proper tool and how to keep it in working condition.
3. Carpenters are expected to have their own hand tools and keep them in good working condition.

OBJECTIVES

- Identify and describe the use of hand tools that are most commonly used by carpenters.
- Use hand tools in a safe and appropriate manner.
- Maintain hand tools in suitable working condition.

Glossary of Hand Tool Terms

Crosscut – a cut made across the grain of lumber

Dado – a cut, partway through, and across the grain of lumber

Groove – a cut, partway through, and running with the grain of lumber

Heel – the back end of objects, such as a handsaw or hand plane

Kerf – the width of a cut made with a saw

Level – horizontal; perpendicular to the force of gravity

Plumb – vertical aligned with the force of gravity

Plumb bob – a pointed weight attached to the line for testing plumb

Rabbet – an L-shaped cutout along the edge or end of lumber

Square – a tool used to mark a layout and mark angles, particularly 90-degree angles; a term used to describe when two lines or sides meet at a 90-degree angle; the amount of roof covering that will cover 100 square feet of roof area

Toe – the forward end of tools, such as a handsaw and hand plane

Whet – honing a tool by rubbing the tool's blade on a flat sharpening stone

VIDEO #1

Explain breakdown of an inch. Explain inch measurements and foot and inch measurements. Have students practice measuring.

A. **Layout Tools**
   1. These are the tools a carpenter uses to measure distances and lay out or mark out lines and angles and set vertical or level pieces.
   2. Measuring Tools.

SLIDES 1 & 2

a. The ability to take measurements quickly and accurately must be mastered early in the carpenter's training.

b. Rules and tapes used in construction in the United States are divided into feet, inches, and sixteenths of an inch (figure 1-1).

c. Most carpenters use the pocket tape, which is commonly available in 6-foot to 30-foot lengths (figure 1-2).

CAUTION Steel tapes can cut the skin when being retracted. They are also easily broken if tape is folded and creased.

d. Steel tapes of 50-foot and 100-foot lengths are commonly used to lay out longer measurements (figure 1-3).

e. Keep steel tapes dry or lightly oiled to prevent rusting.

3. Squares

Reiterate the definition of the term "square" from the glossary of hand-tool terms.

a. The carpenter has the use of a number of different kinds of squares to lay out square and other angle cuts.

Demonstrate the use of a combination square.

b. Combination squares consists of a movable blade, 1 inch wide and 12 inches long, which slides along the body (figure 1-4).
   - A combination square is used to mark or test 90-degree and 45-degree angles.
   - A combination square can function as a depth gauge to lay out or test the depth of rabbets, grooves, and dados (figure 1-5).
   - It can also be used with a pencil as a marking gauge to draw lines parallel to the edge of a board.

VIDEO #3

Demonstrate the use of a speed square.

SLIDE 3

c. Speed squares have a triangular shape, are made of plastic or aluminum alloy, and are available in different sizes, usually 8 or 12 inches (figure 1-6).
   - They are used to lay out angles, especially 90-degree and 45-degree angles.
   - A degree scale on the square allows various angles to be laid out.
   - Other scales on the square are used to lay out rafters.

VIDEO #8

Describe and/or demonstrate that the 1-1/2 inch width of the tongue is the same width as all 2× dimension lumber, which aids in indicating the placement/layout of various framing members.

d. Framing squares, often called steel squares (figure 1-7), are L-shaped and made of steel or aluminum.
   - The longer of the two legs is called the blade or body and is 2 inches wide and 24 inches long.
   - The shorter leg is called the tongue and it is 1 1/2 inches wide and 16 inches long.
   - The outside corner is called the heel.
   - A number of different tables are stamped on both sides of the square. These tables are used to lay out rafters and determine diagonal measurements.
   - Framing squares are also used to lay out stairs and bridging.

Various angles may be set or determined by using the Sliding T-Bevel in combination with a power miter saw as a protractor of sorts.

e. Sliding T-bevels are sometimes called bevel squares (figure 1-8) and are used to lay out or test angles other than those laid out with squares.

4. Levels

There are single vial and double vial levels. When reading a double vial level, always read the top vial.

a. The term level is used to indicate that which is horizontal, and the term plumb is used to indicate that which is vertical (figure 1-9).

VIDEO #3
VIDEO #5

b. Carpenter's level is used to test both level and plumb surfaces (figure 1-10).
   - The air bubble in the vial must be exactly centered between the lines marked on the tube.
   - Caution: Care must be taken not to drop the level as it may break the vial and/or affect accuracy.
   - Test for accuracy (figure 1-11).

SLIDE 4

The use of a braided line such as a mason's line will greatly reduce spinning of the plumb bob.

c. Plumb bob is suspended from a string and hangs absolutely vertical when it stops swinging (figure 1-12).

SLIDE 5

- It can be suspended from a great height to establish one point that is plumb over another (figure 1-13).

5. Chalk Lines

VIDEO #3

   a. A line coated with chalk dust is stretched tightly between two points and snapped near the center against a flat surface (figure 1-14).
   b. Chalk lines consist of a line and reel filled with chalk dust (figure 1-15).
   c. Chalk is usually red or blue. Red tends to be permanent and will bleed through many paints.
   d. Using the chalk line:

Snap lines to shop floor using Pythagorean's Theorem/3—4—5 to make a square corner.

      - When unwinding the chalk line, keep it off the surface until the line is ready to be snapped.
      - Make sure the line is stretched tight before snapping in order to produce a neat and straight line.
      - If the line sags, take the sag out by supporting the line near the center.
      - Keep the line from getting wet.

VIDEO #6

6. Scribers

VIDEO #8

   a. Scribers (figure 1-16) are similar but have a pencil attached to one end.
   b. Wing dividers (figure 1-17) can be used as a compass or to space off equal distances.
   c. Scribing is a technique of laying out stock to fit against irregular surfaces.

Never carry chisels in pockets for safety reasons.

B. Edge-cutting Tools

1. Wood Chisel

Demonstrate the proper technique of sharpening chisels. Have students sharpen a chisel.

   a. Wood chisels (figure 1-18) are used to cut recesses in wood for such things as door hinges and locksets.
   b. Chisels come in widths of 1/4 inch, 1/2 inch, 3/4 inch, 1 inch, and 1 1/2 inches.
   c. Be sure to keep both hands behind the cutting edge (figure 1-19).
2. Bench Planes
   a. Bench planes (figure 1-20) are used for smoothing rough surfaces.
   b. Bench planes are given names according to their lengths. The longest is called the jointer, then the fore, jack, and smooth planes.
3. Block Planes

Have students plane the edge of two boards in preparation for lamination.

Demonstrate proper sharpening techniques.

   a. Block planes (figure 1-21) are smaller than bench planes and are designed to be held in one hand.
   b. The bevel of the cutting edge is on the top side in block planes. On bench planes, it is on the bottom (figure 1-22).
   c. Refer to figure 1-23 for proper use of the block plane.
   d. To produce a keen edge, chisels and plane irons must be whetted (sharpened) using an oilstone or water stone (figure 1-24).
4. Tin Snips

VIDEO #6

   a. Tin snips (figure 1-25a) are also referred to as aviation snips.
   b. Tin snips are used to cut thin metal such as roof flashing and plastic such as vinyl siding.

VIDEO #5

   c. Three styles of aviation snips are available (figure 1-25). Yellow handles are for straight cuts, green handles are for cutting curves to the right, and red handles are for cutting curves to the left.
5. Utility Knife

VIDEO #7

   a. The utility knife (figure 1-26) has replaceable blades.

CAUTION The utility knife can be considered the most dangerous tool in the shop. Always keep your free hand behind the blade and never cut toward your body.

b. It is used to cut a wide variety of material such as gypsum board, soft boards, asphalt shingles, etc.

C. Tooth Cutting Tools

1. Handsaws (figure 1-27).

VIDEO #8

Have students demonstrate use of handsaws.

   a. Crosscut saws cut across the grain.
   b. Ripsaws cut with the grain.
   c. Crosscut saws have teeth shaped like a knife. Ripsaws have teeth shaped like chisels (figure 1-28).
   d. The teeth of handsaws have "set" to prevent the saw from binding.
   e. Kerf is the width of the cut made in the wood by the saw.
   f. Handsaws are designed to cut on the down stroke (figure 1-29).
2. Compass and Keyhole Saws
   a. Compass saws (figure 1-30) are used to make curved cuts.
   b. Keyhole saws are similar to the compass saw except that the blade is narrower.
   c. In soft material, such as gypsum board, the point of the saw can be simply pushed through the material to start.

VIDEO #7

Demonstrate and have students cut a coped joint.

3. Coping Saw
   a. Coping saws (figure 1-31) are used to make a coped joint.
   b. Blades may be installed pointing away or pointing towards the handle.

A minimum of two teeth per thickness of material is recommended.

4. Hacksaws
   a. Hacksaws (figure 1-32) are used to cut various materials but usually metal. Coarse tooth blades are used for thicker material and finer blades are used on thinner material.
5. Wallboard Saw
   a. The wallboard saw (figure 1-33) is a compass saw especially designed for gypsum board.

D. Fastening Tools

VIDEO #3

1. Hammers
   a. Claws may be straight or curved.
   b. Sixteen- to 20-ounce hammers are for finish work and have handles of approximately 12″ in length (figure 1-34).
   c. Twenty- to 30-ounce hammers are used for framing and have handles of 14–18″. Longer Handle (figure 1-35).

VIDEO #6

2. Nail Sets
   a. Nail sets (figure 1-36) are used to set nail heads below the surface.
   b. Most common sizes: 1/32, 2/32, 3/32. Size refers to diameter of the tip.
3. Nailing Techniques
   a. Face nailing describes nailing straight through one piece into another.
   b. Toe nailing is a technique of driving nails at an angle to fasten the end of one piece to another (figure 1-39).
   c. Blunting the nail point helps prevent splitting the wood (figure 1-40).
   d. Drive nails at an angle for greater holding power (figures 1-41 and 1-42).
4. Screwdrivers
   a. Slotted screwdrivers have straight tips of different sizes (figures 1-43a and 1-44).

b. Phillips screwdrivers have cross-shaped tips (figure 1-43b). Phillips come in size Number 1, Number 2, and Number 3.

c. Other screwdrivers include the Robertson screwdriver, which has a squared tip.

Demonstrate the use of clearance holes and pilot holes when driving screws.

d. Screwdriver bits (figure 1-45) are designed to drive a screw using a drill or screw gun.

e. Select screws so that 2/3 of their length penetrates the piece that they are gripping.

VIDEO #2

E. Dismantling Tools

1. Hammers

With the nail in the claw, rocking the hammer from side to side, also is an effective way of pulling a nail. Have students practice face nailing and toe nailing.

a. Claw hammers are used for pulling nails.

b. To increase leverage, place a small block of wood under the hammer head (figure 1-46).

2. Bars and Pullers

a. Wrecking bars (figure 1-47) are available in lengths from 12 to 36 inches.

Demonstrate proper use of all bars and pullers available in shop.

b. A flat bar (figure 1-48a) is used to pry small work and pull small nails.

c. A nail claw, also called a cat's paw (figure 1-50), is used to extract nails that have been driven all the way in.

3. Holding Tools

a. An adjustable wrench (figure 1-51) is used to turn nuts, lag screws, and bolts.

Demonstrate proper use of all clamps available in shop.

b. Combination pliers (figure 1-52) are designed for general use such as extracting, turning, or holding objects.

c. C clamps (figure 1-53) are used for holding objects together. The size is determined by the throat opening.

d. Spring clamps open like scissors (figure 1-53).

e. Wood screws (figure 1-53) are hardwood blocks with threaded rods through them.

f. Quick clamps (figure 1-53) are designed for speed and ease of operation. They are one of many types of bar clamp.

Chapter 2 Power Tools

Instructor's Note

In this column, instructors will find a variety of lecture tips to aid student comprehension, notes on when to use PowerPoint slides to teach concepts, safety cautions for your students, and notes on when your class should view a video segment.

Appendix A lists corresponding numbered PowerPoint Presentation titles (see page 197)

Appendix B lists corresponding numbered Video titles and sections as applicable (see page 201)

Introduction

1. Power tools have been created to increase the productivity of most jobsite tasks.
2. However, with increased speed and production comes increased personal risk.
3. Safe operation of power tools requires knowledge and discipline.
4. Learn the safe operation techniques from the manufacturer's recommended instructions before operating any tool.
5. Eye and ear protection are required.
6. Be sure to emphasize electrical precautions.

OBJECTIVES

- State general safety rules for operating power tools.
- Describe and safely use the following: circular saw, saber saw, reciprocating saw, drill, hammer-drill, screwdriver, plane, router, sander, stapler, nailer, powder-actuated driver, table saw, and power miter saw.

Glossary of Power Tool Terms

Bevel – the sloping edge or side of a piece at any angle other than a right angle

Chamfer – an edge or end bevel that does not go all the way across the edge or end

Compound miter – a bevel cut across the width and also through the thickness of a piece

Crosscut – a cut made across the grain of lumber

Fence – a guide for ripping lumber on a table saw

Miter – cutting the end of a piece at any angle other than a right angle

Miter gauge – a guide used on the table saw for making miters and square ends

Rip – sawing lumber in the direction of the grain

A. Portable Power Tools

1. General Safety Rules

VIDEO #3

   a. Have a complete understanding of a tool before attempting to operate it.
   b. Electrical shocks have caused many fatal accidents. Make sure tools are properly grounded and connected to a ground fault circuit interrupter (GFCI).
   c. Always wear eye and ear protection when operating power tools.
   d. Never use a power tool with a frayed or worn-out cord.

VIDEO #6
VIDEO #7

   e. Use extension cords that are the proper wire size.
   f. Place extension cords out of the way of jobsite traffic.

g. Always unplug tools when making adjustments to the cutter or blades.
h. Watch for loose clothing that might become caught in a tool.
i. Secure the material being worked.
j. Make certain blades are sharp.
k. Use the safety guards as designed by the tool manufacturer.
l. Do not allow your attention to be distracted.

2. Circular Saws

Use Power Tools Assessment Sheet provided to indicate students' competency of each specific operation involved in the safe and proper use and maintenance of this particular power tool.

a. Sometimes called the Skilsaw, this is the most used portable power tool (figure 2-1).
b. The saw is adjustable for depth of cut and the base may be tilted for making bevel and miter cuts (figure 2-2).
c. These saws are used to make Compound Miters, which are angled cuts across the width and also through the thickness (figure 2-2).
d. The size of the saw is determined by the diameter of the blade, which ranges from 4-1/2 to 16 inches.
e. The blade may be driven directly by the motor or through a worm gear (figures 2-3a and 2-3b).
f. Circular saw blades are available in a number of different styles.
   - The shape and number of teeth determine the cutting action.
   - Carbide-tipped blades are used more than high-speed steel blades.

VIDEO #3
VIDEO #7

g. Using the Portable Circular Saw:
   - Support and secure stock.
   - Never cut between supports (figure 2-4).
   - Adjust depth of cut so blade does not travel more than 1/8th of an inch through stock (figure 2-5).
   - Make sure retractable guard operates properly (figure 2-6).
   - Never wedge guard back in open position.
   - Hold the saw with two hands.
   - Let saw reach full speed before advancing saw into work.
   - Cut on waste side of the mark.
   - Keep the saw clear of your body until the saw has stopped.
   - Let the waste piece drop clear.
   - Sometimes when cutting an angle, it may be necessary to retract the guard before beginning the cut. Release the handle after the cut has been started.

CAUTION Any deviation from the line may cause the saw to bind or kick back. Do not force the saw forward. In case the saw does bind, stop the motor and bring the saw back to where it will run free. Continue following the line closely.

h. Making Plunge Cuts:
   - Plunge cuts are so called because they are used to plunge into material when internal cuts are necessary to make openings for floors, walls, and countertops (figure 2-7).
   - Lay out the cut to be made. Adjust the saw for the depth to be cut.
   - Hold the guard open and tilt the saw up with the front edge of the base resting on the work.
   - Make sure the teeth of the blade are not touching the work.
   - Start the saw, and lower the blade slowly into the work.
   - Before moving the saw from the cut, release the switch and wait for saw to stop. Caution: Do not move the saw backwards.

Use Power Tools Assessment Sheet provided to indicate students' competency of each specific operation involved in the safe and proper use and maintenance of this particular power tool.

3. Saber Saws

a. Saber saws (figure 2-8) are sometimes called jigsaws. They are widely used to make curved cuts.

b. The base of the saw may be tilted to make bevel cuts.

c. Blades are available for fine and coarse cutting of wood and metal.

d. Using the Saber Saw:
- Secure the work.
- With the blade clear, pull the trigger.
- With both hands on the saw, make saw cut from the waste side of the line.
- While cutting, hold the base of the saw down firmly on the work.

e. Making Plunge Cuts:
- Tilt the saw up on the forward end of its base with the blade in line and clear of the work (figure 2-9).
- Start the motor, and very gradually and slowly lower the saw until the blade penetrates the work and the base rests firmly on it.
- Caution: Hold the saw firmly to prevent it from jumping when the blade makes contact with the material.
- Caution: Make sure the tool comes to a complete stop before withdrawing it from the material being cut.

4. Reciprocating Saws

Use Power Tools Assessment Sheet provided to indicate students' competency of each specific operation involved in the safe and proper use and maintenance of this particular power tool.

a. Sometimes called the Sawzall, reciprocating saws (figure 2-10) are primarily used for roughing in work.

b. Can cut a variety of materials such as metal, plastic, wood, and nail-embedded wood.

c. Some models have an orbital cutting feature.

d. Blades are available for wood cutting, metal cutting, and nail-embedded wood cutting.

e. Blade lengths run from 4 to 12 inches.

f. Using the reciprocating saw:
- Hold the base or shoe of the saw firmly against the work.
- Pull the trigger and ease the saw into the cut.
- Holes may be pre-drilled for the blade to make cut-outs.

5. Drills and Drivers

Generally speaking, the larger the bit, the slower the speed, and the smaller the bit, the faster the speed.

a. Manufactured in a number of styles and sizes.
- Electric pistol grip (figure 2-11a).
- Electric D-shaped handle (figure 2-11b).
- Rechargeable battery-operated (figure 2-18).

b. The size of the drill is determined by the capacity of the chuck. Most popular are 1/4 inch, 3/8 inch, and 1/2 inch (figure 2-12).

c. Most drills have variable speeds and reversible controls.

d. Bits
- Twist drill bits range in size from 1/16 inch to 1 inch, in increments of 1/64 inch.
- Auger bits (figure 2-13) range from 1/4 inch to 1 inch, in 1/16-inch increments.
  1) The bit size is designated by the number of 1/16-inch increments in its diameter. For instance, a Number 12 bit is 12/16, and therefore will bore a 3/4-inch diameter hole.
  2) Auger bits are available with coarse or fine feed screws.
- Expansive bits (figure 2-14) are for holes over 1 inch and up to 3 inches.

- Speed bits (figure 2-15).
  1) Sometimes called spade bits.
  2) Sizes range from 1/4 inch to 2 inches in diameter.

Never use bits with threaded center points in electric drills when boring deep holes.

- Power bores (figure 2-15) are for making a hole with a cleaner edge in finish work.
- Hole saws are used to cut holes through thin material. Sizes range from 5/8 of an inch to 6 inches in diameter.
- Masonry drill bits have a carbide tip for drilling holes in concrete, brick, tile, and other masonry. They are most efficiently used in hammer drills.

e. Using Portable Electric Drills:

- Select proper size of bit and drill.
- Insert drill and tighten chuck with chuck key or by holding the chuck of a keyless chuck.

VIDEO #1

- Secure stock.
- Start motor and apply pressure as required.
- Caution: Remove bit from the hole frequently while drilling to remove chips. Be ready to release trigger switch if the drill bit binds.

f. Hammer-drills (figure 2-16) are similar to other drills. They deliver as much as 50,000 hammer blows per minute. They are most commonly 3/8 and 1/2 inch in size.

g. Screwguns (figure 2-17) are used extensively for fastening gypsum board.

- Chuck is made to receive screwdriver bits.
- Clutch-operated.
- Adjustable nosepiece used to set depth of screw.

6. Portable Power Planes

Use Power Tools Assessment Sheet provided to indicate students' competency of each specific operation involved in the safe and proper use and maintenance of this particular power tool.

a. Used primarily to smooth and straighten long edges (figure 2-19).

b. Caution: Extreme care must be taken when operating as there is no retractable guard and the high-speed cutter head is exposed on the bottom of the plane. Keep the tool clear of your body and power cords until the cutter head has completely stopped.

c. Operating Power Planes

- Secure work being planed.
- Adjust for depth of cut.
- Apply pressure to the toe of the plane at the beginning of the cut.
- Apply even pressure after the heel (back) is on the piece.
- Move pressure to the heel (back) at the end of the cut.

7. Routers

Use Power Tools Assessment Sheet provided to indicate students' competency of each specific operation involved in the safe and proper use and maintenance of this particular power tool.

a. Routers (figure 2-20) are available in 1/4 horsepower to 3 horsepower with speeds from 18,000 to 30,000 rpm.

b. Used to smooth cut edges with a variety of shapes and sizes of bits (figure 2-21).

c. Adjustable base is provided to control depth of cut.

d. Stalling the movement of the router will cause the bit to burn or melt the material.

VIDEO #7

e. Light-duty routers are called laminate routers, used for trimming plastic laminates (figure 2-22).

f. Using the Router:

- Secure stock.
- Select correct bit.

- Insert bit into chuck. Make sure chuck grabs at least twice the shaft diameter of the bit.
- Adjust depth of cut.
- Start motor.
- Advance bit into cut in a direction that is against the rotation of the bit.
- Caution: Keep the router clear of your body and power cords until the bit has stopped completely.

8. Sanders

Use Power Tools Assessment Sheet provided to indicate students' competency of each specific operation involved in the safe and proper use and maintenance of this particular power tool.

a. Belt Sanders (figure 2-23)
- The size of the belt determines the size of the sander.
- Belt widths range from 2 1/2 to 4 inches, and lengths vary from 16 to 24 inches.
- Belts are installed by retracting the forward roller of the belt sander, and installing the belt over the rollers (figure 2-24).
- Turn the adjusting screw one way or the other to adjust the belt and center it on the roller (figure 2-25).

Ensure that trigger is in the unlocked position before plugging in the belt sander.

- Using the Belt Sander:
  1) Secure work to be sanded.
  2) Select appropriate belt grit.
  3) Check that belt is tracking properly.
  4) Hold with both hands and start the machine.
  5) Sand material in parallel strikes with the grain.
  6) Sand with the pad flat on the work.
  7) Do not sand in one spot; always keep the sander moving.
  8) Let the belt stop before setting the tool down.
  9) Caution: Be careful to keep the electrical cord clear of the tool as the cord may be easily get tangled in the sander if the operator is not alert.

VIDEO #7

b. Finishing Sanders

Use Power Tools Assessment Sheet provided to indicate students' competency of each specific operation involved in the safe and proper use and maintenance of this particular power tool.

- Finishing sander (figure 2-26) is used for final sanding of interior work.
- Available in cordless models.
- Motion can be orbital, oscillating, or a combination of both.
- Random orbital sander randomly moves the center of the sanding paper at a high speed.
- Using the Finishing Sander:
  1) Select the desired grit sand paper.
  2) Attach sand paper to pad.
  3) Start the motor and sand surface evenly and slowly.
  4) Do not use excessive pressure.
  5) Always hold the sander flat on its pad.

VIDEO #3

9. Pneumatic Staplers and Nailers (figure 2-27a and 2-27b)

Use Power Tools Assessment Sheet provided to indicate students' competency of each specific operation involved in the safe and proper use and maintenance of this particular power tool.

a. Commonly called guns.

b. Caution: Can be extremely dangerous.

c. Nailing guns
- Framing gun (figure 2-28a) drives smooth shank and ring shank nails in lengths of 1 3/4 inches to 3 1/2 inches. It is primarily used to fasten framing, sub-floor, sheathing, and similar components of a building.
- Finish nailer (figure 2-31a) drives finish nails in lengths from 1 inch to 2 1/2 inches long (figure 2-31b). Depth of nail immersion can be adjusted.

Used for applications of practically all kinds in interior and exterior finish work.

- Brad nailer (figure 2-32a) drives brads ranging in length from 1/2 inch to 1 5/8 inches (figure 2-32b). Used to fasten small moldings and trim.
- Roofing nailer (figure 2-33a) is designed for fastening asphalt and roofing shingles. It drives five different sizes of roofing nails from 7/8 inch to 1 3/4 inches (figure 2-33b).

d. Staplers

- Most common is the roofing stapler (figure 2-35a) that drives 1-inch wide crown staples in lengths of 3/4 inch to 1 1/2 inches (figure 2-35b).
- Staples come glued together in strips (figure 2-36).

e. Cordless Guns

- Pneumatic staplers and nailers are powered by compressed air (figure 2-37).
- Cordless nailers utilize a disposable fuel cell (figure 2-38).

CAUTION Be sure to remove your finger from trigger when not driving fasteners. Never point a pneumatic tool toward yourself or anyone else. Ensure that your material is secure and there is nobody behind or below your work. Always treat every pneumatic tool as though it is loaded. Never leave an unattended gun with the air supply connected.

f. Using Staplers and Nailers:

- Oil gun when required.
- Load magazine with desired fasteners.
- Adjust air pressure when required.
- Press the nose of the gun to the surface of the material and then pull the trigger.
- Pressing the trigger and tapping the nose of the gun to the material will also drive a fastener. This is called "bump firing."

10. Powder-Actuated Drivers

a. Powder-actuated drivers (figure 2-39) are used to drive specially designed pins into masonry or steel.

b. Various strength of powder charges are available.

VIDEO #6
Section 5

c. Drive pins are available in a variety of sizes.

d. Powder charges are color-coded according to strength.

e. Many states require a certification for the operator.

f. Caution: Powder-actuated tools operate the same as a gun. Treat this tool as you would a gun.

g. Using Powder-Actuated Drivers:

- Study manufacturer's direction for safe and proper use of the gun.
- Wear proper ear and eye protection.
- Make sure drive pin will not penetrate completely through the material.
- Make sure the recommended shield is located on the nose of the gun.
- Select the proper fastener.
- Select the proper powder charge. Caution: Always start with the weakest charge that will do the job.
- Load the driver with the pin first and the cartridge second.
- Press the tool hard against the work surface and pull the trigger.
- Caution: If the gun does not fire, hold it against the work surface for at least thirty seconds. If the cartridge is stuck, do not attempt to pry it out with a knife or screw driver, as most cartridges are rim-fired and could explode.

VIDEO #7

B. Stationary Power Tools

1. General Safety Rules

a. Be trained and competent in the use of stationary power tools before attempting to operate them without supervision.

b. Make sure power is disconnected when making adjustments to machines.

c. Make sure saw blades are sharp and suitable for the operation. Ensure that safety guards are in place and that all guides are in proper alignment and secured.

d. Wear eye protection and appropriate, properly-fitted clothing.

e. Keep the work area clear of scrap that might present a tripping hazard.

f. Keep stock clear of saw blades before starting a machine.

g. Do not allow your attention to be distracted while operating power tools.

h. Turn off the power and make sure a machine has stopped before leaving the area. Safety precautions applying to specific operations are given when those operations are described.

VIDEO #7

Use Power Tools Assessment Sheet provided to indicate students' competency of each specific operation involved in the safe and proper use and maintenance of this particular power tool.

2. Table Saws

a. The size of the table saw (figure 2-40) is determined by the diameter of the blade. Most common is the 10-inch model.

b. The blade is adjusted for depth of cut and can be tilted up to 45 degrees.

c. Rip fence guides the work for ripping operations (figure 2-41).

d. Guards should always be in place to protect the operator.

e. Caution: Never reach over the blade, and never stand in back of the saw blade. Make sure nobody else is behind the saw blade in case of a kickback. A splitter and anti-kickback device should be used when the cutting operation allows it.

f. Ripping operations:

- Measure from the rip fence to the saw blade.
- Lock the fence in place.
- Adjust the height of the blade to 1/4 inch above the stock to be cut.
- Adjust and prepare blade guards for use.
- Turn on the power.
- Hold stock against the fence.
- Feed work through saw blade until cut is finished, ensuring it is pushed through clear of the blade.
- If the stock is of sufficient width (at least 5 inches wide) the operator may use his hand. Otherwise a push stick (figure 2-42a) must be used.
- Caution: Make sure the stock is pushed all the way through the saw blade to prevent kickback. Always use a push stick when ripping narrow pieces. Do not pick small waste pieces from the saw table when the saw is running. Never make free-hand cuts (i.e., no rip fence). Never reach over a running saw blade.

g. Crosscutting Operations:

- For most crosscutting operations a miter gauge is used (figure 2-43). It may be locked and turned in any position up to 45 degrees.
- Caution: The miter gauge should not be used at the same time as the rip fence.

VIDEO #7
Sections 2–4

Use Power Tools Assessment Sheet provided to indicate students' competency of each specific operation involved in the safe and proper use and maintenance of this particular power tool.

3. Miter Saws

a. Miter saw is also called the power miter box, with sizes ranging from 8 1/2″ to 12 inches (figure 2-45).

b. Saw blade may be pivoted 45 degrees to the right or left.

c. Compound miter saw will also let the blade tilt left or right 45 degrees to make compound miter cuts (figure 2-46).

4. Band Saws

a. Band saw is a tool designed to cut irregular shapes in various thicknesses of material (figure 2-47).

b. Operated freehand, moving the stock to follow curved lines.

   c. Well suited to resaw thick boards into thinner ones (figure 2-48).
   d. Have two large diameter wheels, one above the table and one below (figure 2-49).
   e. The upper guide should be positioned about 1/4″ above the stock being cut (figure 2-50).
   f. Sizes are determined by the diameter of the drive wheels.
   g. Caution: Take care to avoid backing out of long cuts. Backing out could cause the band to come off the drive wheels.
   h. Width of the band affects the turning radius on the cut (figure 2-51).
5. Jointers
   a. Used to smooth stock surfaces to remove the marks made by sawing (figure 2-52).
   b. Size is determined by width of the table.
   c. Spring-loaded safety shield swings in place to cover the portion of the cutter head not being used.
   d. Constructed with two tables—infeed and outfeed, always kept parallel to each other. Offset height of the tables determines the depth of cut (figure 2-53).
   e. Infeed table is adjustable in height. Outfeed table is stationary and set at a height equal to the extreme cutting height of the cutting head.
   f. Caution: Check manufacturers recommendations for the maximum allowable depth of cut.
   g. Safe operation requires two hands to hold stock securely. Begin with the stock resting on the infeed table and fence. Concentrate on keeping the stock against the table and fence during the entire pass without it wobbling. Then slowly ease the stock towards the safety shield and into the cutter head (figure 2-54).
   h. Caution: Avoid using fingers to hold the stock directly over the cutters.
   i. Other operations of the jointer include chamfers and rabbets (figure 2-55).
6. Drill Press
   a. Precision drilling tool consisting of a base, column, table, and head (figure 2-56).
   b. Bit speed is adjustable by typically changing the belt location on the cone-shaped pulleys.
   c. Caution: Take care to secure the work to the table before drilling. If the bit binds in the hole, the entire piece may spin out of control.
7. Shaper
   a. A shaper used to make molding and decorative edges in irregular surfaces (figure 2-57). It is a heavy-duty router attached upside down to a base.
   b. The stock may be tooled using the fence as a guide or using a collar on the bit.
   c. Collar may be located on the bit itself, which rides along any irregular shape.
   d. Caution: Great care must be taken when setting up the shaper. Check the cutters carefully to be sure they are sharp and in proper working condition. If a tip should break off while the cutter is spinning, it could result in serious injury.
8. Bench Sander
   a. Bench sanders in the form of a drum, disk, and belt are designed to smooth most any shaped piece of work (figure 2-58). The material is moved over the sander to smooth and remove material fast.

Chapter 3 Wood and Wood Products

Instructor's Note

In this column, you will find a variety of lecture tips to aid student comprehension, notes on when to use PowerPoint slides to teach concepts, safety cautions for your students, and notes on when your class should view a video segment.

Appendix A lists corresponding numbered PowerPoint Presentation titles (see page 197)

Appendix B lists corresponding numbered Video titles and sections as applicable (see page 201)

Introduction

1. The construction material most often associated with carpentry is wood.
2. Trees are harvested and sawed into lumber of various sizes and shapes, having a variety of characteristics.
3. It is necessary to understand the nature of wood to get the best results from the use of it.
4. The industry of engineered lumber has emerged to reduce manufacturing waste and improve the strength of wood products.

OBJECTIVES

- Define hardwood and softwood, and give examples of some common kinds.
- State the grades and sizes of lumber.
- Measure linear foot, compute square foot and board foot measure.
- Describe the composition, kinds, sizes, and several uses of plywood, oriented strand board, particleboard, hardboard, medium-density fiberboard, and soft board.
- Describe the uses and sizes of laminated veneer lumber, parallel strand lumber, laminated strand lumber, wood I-beams, and glue-laminated beams.

Glossary of Wood and Wood Product Terms

Air dried – technique of removing water from lumber using natural wind currents

Annular ring – the rings seen when viewing a cross-section of a tree trunk; each ring constitutes one year of tree growth

Board – lumber usually less than 2 inches thick

Board foot – a measure of lumber volume that equals 1 foot square and 1 inch thick or any equivalent lumber volume; the letter M is used to represent 1000 board feet

Cambium layer – a layer just inside the bark of a tree where new cells are formed

Coniferous – trees that are cone bearing; also known as evergreen trees

Deciduous – trees that shed leaves each year

Dimension – a term used to define a measurement of an item; also used to refer to all 2 × lumber used in framing

Dry kiln – large ovens used to remove water from lumber

Finger joint – a process where shorter lengths are glued together using deep, thin V grooves resulting in longer lengths

Header – framing members placed at right angles to joists, studs, and rafters to form and support openings

Heartwood – the wood in the inner part of a tree, usually darker and containing inactive cells

Lumber – general term for wood that is cut from a log to form boards, planks, and timbers

Medullary ray – bands of cells radiating from the cambium layer to the pith of a tree to transport nourishment toward the center

Millwork – any wood products that have been manufactured, such as moldings, doors, windows, and stairs, for use in building construction; sometimes called joinery

OC ("on center") – the distance from the center of one structural member to the center of the next one

Panel – a large sheet of building material that usually measures $4' \times 8'$

Pith – the small, soft core at the center of a tree

Plain-sawed – a method of sawing lumber that produces flat-grain in which the annular rings tend to be parallel to the width of the board

Quarter-sawed – a method of sawing lumber that produces a close grain pattern in which the annular rings tend to be perpendicular to the width of the board

Sapwood – the outer part of a tree just beneath the bark containing active cells

Sawyer – a person whose job is to cut logs into lumber

Tempered – treated in a special way to make a material harder and stronger

Timbers – large pieces of lumber over 5 inches in thickness and width

VIDEO #3

A. Wood

1. Wood
   a. Wood can be cut, shaped, or bent in just about any form.
   b. There are many kinds of wood that vary in strength, workability, elasticity, color, grain, and texture.
   c. Wood is one of our greatest natural renewable resources.
   d. Heartwood is the central part of the tree and usually darker in color and more durable than sapwood. The heartwood of cedar, cypress, and redwood, for instance, is extremely resistant to decay.
   e. Sapwood is between the heartwood and the bark of the tree and is usually lighter in color.
   f. Annual rings determine the age of the tree (figure 3-1).
2. Hardwood and Softwood

Take students on a walk to identify various types of trees; have samples of local manufactured lumber from the trees you are viewing.

   a. Hardwood comes from deciduous trees that shed their leaves each year.
   b. Softwood is cut from coniferous, or cone-bearing trees, commonly known as evergreens (figure 3-2).
   c. Generally speaking, softwood is softer than hardwoods; however, some softwoods can be harder than some hardwoods. Example: Fir, a softwood, is harder and stronger than basswood, a hardwood.
   d. The best way to learn the different kinds of wood is by working with them.
3. Manufacture of Lumber
   a. Logs arrive at the sawmill and their bark is removed.
   b. Huge saws slice the log into large planks and ultimately result in various lumber.
   c. Lumber is either air dried or kiln dried (figures 3-3 and 3-4).
   d. Once dry, the rough lumber is surfaced to standard sizes and shipped.
   e. Refer to figure 3-5 for the specific names of lumber surfaces.

f. Plain-sawed lumber
  - This method of sawing is the least expensive and produces the greatest width (figures 3-6a and 3-6b).
  - Plain-sawed lumber shrinks more when drying and warps easily.

g. Quarter-sawed lumber
  - This method produces pieces in which the annular rings are at or almost at a right angle to the wide surface (figures 3-7a and 3-7b).
  - This lumber has less of a tendency to warp and shrinks less and more evenly when it dries.
  - The wood often has a more appealing grain pattern.

h. Combination-sawed lumber is a computer-guided method of optimally cutting a log (figure 3-8).

i. Moisture Content and Shrinkage
  - When a tree is freshly cut, it contains a great amount of water.
  - When lumber is first cut from the log, it is called green lumber (figure 3-9).
  - As wood dries, it shrinks and usually warps (figure 3-10).
  - It is important to prevent water from getting behind the wooden surface (figure 3-11).

j. Lumber Storage
  - Have lumber delivered in proper sequence (e.g., those that are to be used first should be situated on top).
  - Stored lumber should be adequately protected from moisture and excessive sunlight.
  - Use short lengths of rough lumber (also known as dunnage) under piles to keep the pile from sagging and coming in contact with the ground.
  - Caution: Do not pile lumber in such a manner that could cause it to topple.

VIDEO #6
Section 3

Show students examples of lumber defects.

k. Lumber Defects
  - Cups, bows, twists, and crooks (also called crowns) are shown in figure 3-12.
  - Knots are cross-sections of branches in the trunk of the tree.
  - Pitch pockets are small cavities that hold pitch, which sometimes oozes out.
  - Wane is bark on the edge of the lumber or the surface from which the bark has fallen.

SLIDE 6

l. Lumber Grades and Sizes
  - Wood product associations establish sizes and grades of lumber.
  - Mills use the association grade stamp to indicate quality and grade standards (figure 3-13).
  - Boards are less than 2 inches thick. The board group is divided into boards, sheathing, and form lumber (figure 3-14).
  - Dimension lumber is 2–4 inches thick. The dimension group is divided into framing, studs, structural light framing, and structural joists and planks (figure 3-14).
  - Timbers are 5 inches and thicker. Timbers are divided into beams and stringers (figure 3-14).
  - Hardwood Lumber
    1) Select and first and seconds (FAS) are the best grades of hardwood.
    2) The length and width of the board determines whether it is select or FAS.
    3) The less restrictive grades, in decreasing order, are Number 1 common, Number 2 common, Number 3A, and Number 3B.

- Rough lumber that comes directly from the sawmill is close in size to what is called nominal size.
- When rough lumber is planed, it is reduced in thickness and width to standard and uniform sizes. Therefore, when dressed (surfaced), although a piece of wood may be called a 2 × 4, its actual size is 1 1/2 inches by 3 1/2 inches (figure 3-15).

Use the following rule to determine actual lumber sizes. For nominal dimensions under 2 inches, subtract 1/4 inch to determine actual size. For nominal sizes between 2 and 6 inches, subtract 1/2 inch to determine actual size. For nominal sizes between 8 and 12 inches, subtract 3/4 inch to determine actual size.

4. Board Measure
   a. Softwood lumber is usually purchased by specifying the number of pieces. Example: 35/2 × 6 × 16.
   b. When no particular lengths are required, the total number of linear feet is ordered.
   c. When bought in large quantities, lumber is priced and sold by the board foot. A board foot is a measure of the volume of lumber. It is equivalent to a piece 1 inch thick, 12 inches wide, and 1 foot long (figure 3-16).
   d. To calculate the number of board feet, use the following formula: number of pieces × thickness in inches × width in inches × length in feet divided by 12 = number of board feet.

Have students calculate number of board feet with various lumber examples.

B. Engineered Panels

VIDEO #3
Section 4

1. The term engineered panels refers to man-made products in the form of large reconstituted sheets, sometimes called panels or boards.
2. Plywood
   a. The American Plywood Association (APA) inspects, tests, and certifies panels to government and industrial standards.
   b. Grade stamp assures the consumer that the product has met quality and performance requirements (figure 3-17).
   c. The most extensively engineered panel is plywood (figure 3-18).
3. Veneer Grades
   a. A, B, C plugged, C, and D are used to indicate the appearance quality of panel veneers.
   b. Two letters are found in the grade stamp. One letter indicates the quality of the face, and the other indicates the quality of the opposite face.
4. Strength Grades
   a. Used to determine the strength of the panel.
   b. Group 1 is the strongest and most often used.

VIDEO #3
Section 5

Provide sample of OSB.

5. Oriented Strand Board (OSB)
   a. OSB is composed of small oriented strand-like wood pieces arranged in three to five layers with each layer perpendicular to the previous layer (figure 3.19).

Provide examples of stamps.

6. Performance Ratings (figure 3-20)
   a. Designates panels' use, and include the following:
      - APA-rated Sturd-I-Floor intended for single layer flooring.
      - APA-rated Siding, rated for siding.
      - APA-rated Sheathing, used for roofs, floors, and walls.
      - Structural 1, intended for special high-strength applications.
7. Exposure Durability (there are three ratings)
   a. Exterior – designed for permanent exposure to weather or moisture.
   b. Exposure 1 – intended to withstand exposure to weather for long periods of time.
   c. Exposure 2 – designed for use where only moderate exposure to weather is expected.

8. Span Ratings
   a. Appears as two numbers separated by a slash, such as 32/16.
   b. Left number denotes maximum spacing of supports when the panel is used for roof or wall sheathing. Right number indicates the maximum recommended spacing used for supports when panel is used for sub-flooring.

C. Non-Structural Panels

1. Hardwood Plywood
   a. Hardwood faced veneers, of which the most popular are birch, oak, and lauan.
   b. Interior use such as paneling, cabinets, etc.
2. Particleboard
   a. Made of wood flakes, chips, sawdust, shavings, and adhesive (figure 3-21).
   b. Quality of particleboard is indicated by its density.
   c. Used for construction of kitchen cabinets and countertops.
3. Fiberboard
   a. Manufactured as high-density, medium-density, and low-density.
   b. Hardboard (or high-density board) as the most popular. Hardboard may be used inside or outside. Common uses include cabinet backs, drawer bottoms, and siding.
   c. Medium-density fiberboard (MDF) is for interior use only. Used for case goods, drawer parts, kitchen cabinets, cabinet doors, signs, and some interior wall finish.
   d. Low-density fiberboard is also called softboard (figure 3-22). Primarily used for insulating or sound-control purposes.

D. Engineered Lumber

The three main advantages to engineered lumber are its strength, the wide availability of sizes (particularly in length), and the decreased time of assembly in comparison to dimension lumber.

1. Engineered lumber products are reconstituted wood products and assemblies designed to replace traditional structural lumber (figure 3-23).
2. Laminated Veneer Lumber (LVL) (figure 3-25)
   a. Intended for use as high-strength, load-carrying beams.
   b. Can be cut with regular tools and requires no special fasteners (figure 3-24).
3. Parallel Strand Lumber (PSL)
   a. Often known by its brand name, Parallam (figure 3-26).
   b. Used primarily for beams or posts.
4. Laminated Strand Lumber (LSL)
   a. Commonly known by its brand name, Timberstrand (figure 3-27).
   b. Used for a wide range of millwork, such as doors, windows, and virtually any product that requires high-grade lumber. Also used for window and door headers.
5. Wood I-Beams (figure 3-28)
   a. Engineered wood assemblies that utilize an efficient I shape, common in steel beams, which gives them tremendous strength in relation to their size and weight.
   b. Flanges of the beam may be laminated veneer lumber, finger-jointed solid wood lumber, or solid lumber (figure 3-29).
   c. The web of the beam may be made of plywood, laminated veneer lumber, or oriented strand board (OSB).
   d. Used for floor joists, roof rafters, and door and window headers.
   e. Available in various sizes (figure 3-30).

6. Glue-Laminated Lumber
   a. Commonly called Glulam.
   b. Constructed of solid lumber glued together side against side to make beams and joists of large dimensions that are stronger than natural wood of the same size.
   c. Sometimes left exposed to show the natural wood grain.

Chapter 4 Fasteners

Instructor's Note

In this column, you will find a variety of lecture tips to aid student comprehension, notes on when to use PowerPoint slides to teach concepts, safety cautions for your students, and notes on when your class should view a video segment.

Appendix A lists corresponding numbered PowerPoint Presentation titles (see page 197)

Appendix B lists corresponding numbered Video titles and sections as applicable (see page 201)

Introduction

1. Fasteners come in many styles, shapes, and sizes requiring different fastening techniques.
2. It is important that a carpenter know which fasteners are available and which securing technique should be employed, and to be able to wisely select the most appropriate fastener for various materials under different conditions.

OBJECTIVES

- Name and describe various types of nails and their proper use.
- Name and describe various types of screws and their proper use.
- Name and describe lag screws and their proper use.
- Name and describe various types of bolts and their proper use.
- Name and describe solid wall anchors and their proper use.
- Name and describe hollow wall anchors and their proper use.
- Name and describe a variety of adhesives and their proper use.

Glossary of Fastener Terms

Anchor – a device used to fasten structural members in place

Brad – a thin, short, finishing nail

Duplex nail – a double-headed nail used for temporary fastening such as in the construction of wood scaffolds

Electrolysis – the decomposition of one of two unlike metals in contact with each other in the presence of water

Face nailing – method of driving a nail straight through a surface material into supporting member

Finish nail – a thin nail with a small head designed for settling below the surface of finish material

Galvanized – protected from rusting by a coating of zinc

Mastic – a thick adhesive

Penny (d) – a term used in designating nail sizes

Toenailing – method of driving a nail diagonally through a surface material into supporting member

A. Nails

1. There are hundreds of kinds of nails manufactured for just about any kind of fastening job.

2. Nails may be made of aluminum, brass, copper, steel, and other metals.
3. Some nails are hardened so they can be driven into masonry without bending.
4. Most commonly used nails are depicted in figure 4-1.
5. Uncoated nails are called bright nails.
6. To prevent rusting, steel nails are coated with zinc. These are called galvanized nails.

Have examples of bright, galvanized, stainless steel, and masonry nails.

7. When fastening metal that is going to be exposed to the weather, use nails of the same metal as the material being fastened to prevent electrolysis.
8. Caution: Always wear eye protection when driving nails.
9. Stainless steel nails must be used in cedar, redwood, and oak when exposed to weather to prevent bleeding.

SLIDE 7

B. Nail Sizes

1. Designated by the Penny system
2. The symbol for Penny is d; perhaps it is the abbreviation of Denarius, for an ancient Roman coin.

Have students size nails by Penny size.

3. The Penny size of a nail is determined by its length. Example: 2d is 1 inch long and 6d is 2 inches long (figure 4-2).

C. Kinds of Nails

1. Common Nails
   a. Made of heavy gauge wire.
   b. Have a medium-sized head and a smooth shank (figure 4-3).
2. Box Nails
   a. Similar to common nails except they have a thinner shank.
   b. Often used close to edges and ends because they have less tendency to split the material (figure 4-3).
3. Finish Nails
   a. Light gauge with a small head, used to fasten interior trim.
   b. The head is sunk into the wood with a nail set and filled with a filler (figure 4-3).
4. Casing Nails
   a. Similar to finish nails with the head and shank slightly larger than a finish nail.
   b. Used to fasten exterior trim.
5. Duplex Nails
   a. Have a double head.
   b. Designed to be easily removed (figure 4-4).
6. Brads
   a. Small finish nails from 1/2 inch to 1 1/2 inches in length.
   b. Have a thinner shank diameter.
7. Roofing Nails
   a. Short with a heavy gauge, with wide, round heads.
   b. Used for fastening roofing material and softboard wall material.
   c. Galvanized.
   d. Sizes run from 3/4 inch to 2 inches.
8. Masonry Nails
   a. May be cut nails or wire nails (figure 4-6).

b. Made from hardened steel.
c. Very brittle; may break and shatter.
d. Caution: Always wear safety glasses.

D. Screws

1. Greater holding power than nails.
2. Can be removed.
3. When ordering screws, specify length, gauge, type of head, coating, kind of metal, and screwdriver slot.
4. Screw sizes
   a. Lengths range from 1/4 inch to 6 inches.
   b. Gauges run from 0 to 24. The lower the gauge, the thinner the screw.

E. Kinds of Screws

1. Wood Screw
   a. Identified by the shape of the head and the screwdriver slot (figures 4-7, 4-8, and 4-10).
   b. Come in a variety of metals and coatings.

Show examples of sheet metal screws and self-tapping screws. Demonstrate application of both.

2. Sheet Metal Screws
   a. Used for fastening thin metal.
   b. Includes the self-tapping screw (figure 4-9).
   c. Caution: Care should be taken when screws are driven with drills or screw guns. Metal slivers can be received from spinning screws.

Demonstrate installation of lag screw.

3. Lag Screws (figure 4-11)
   a. Similar to wood screws, except they are larger and have a hex head.
   b. Are sized by diameter and length. Diameters range from 1/4 to 1 inch; lengths range from 1 inch to 12 inches.
   c. Shank and pilot holes to receive lag screws are drilled in the same manner as for wood screws.
   d. Wax the threads to allow screw to turn more easily.
   e. Place a flat washer under the head to prevent the head from digging into the wood.

F. Bolts

1. Bolts are similar to screws except their threads are designed to accept a nut.
2. Kinds of Bolts

Provide examples of different bolts and bolt sizes.

   a. Carriage bolt has an oval head.
   b. Machine bolt has a square or hex head.
   c. Stove bolts have round or flat heads with a screwdriver slot.
3. Bolt Sizes
   a. Bolt sizes are specified by diameter and length.

G. Anchors

Demonstrate use of various anchors as available.

1. Anchors are special kinds of fasteners used to attach parts to solid masonry and hollow walls or ceilings.
2. Solid wall anchors
   a. Heavy-duty anchors
      - Wedge anchor (figures 4-13 and 4-14) is used when high resistance to pull-out is required.

        - Sleeve anchor (figure 4-15) is expanded by tightening the nut.
        - Drop-in anchor (figure 4-16) consists of an expansion shield and a cone-shaped expander plug.
    b. Medium-duty anchors
        - Split fast anchors (figure 4-17) are one-piece steel with two sheared expanded halves at the base.
        - Single and double expansion anchors (figure 4-18) are used with machine screws or bolts.
        - Lag shield (figure 4-19) is used with a lag screw.
        - Concrete screws (figure 4-20) utilize specially fashioned high and low threads that cut into a properly sized hole in concrete.
    c. Light-duty anchors
        - Hammer drive anchor (figure 4-21) has a body of zinc alloy containing a steel expander pin.
        - Aluminum driver anchor has both body and pin made of aluminum to avoid electrolysis reaction.
        - Nylon nail anchor utilizes a nylon body and a threaded steel expander pin.
        - Lead and plastic anchors, also called plastic inserts (figure 4-22) are commonly used for fastening lightweight fixtures to masonry walls.
3. Hollow Wall Fasteners
    a. Toggle bolts (figure 4-23) have a wing or tumble toggle.
    b. Plastic toggles (figure 4-24) consist of four legs attached to a body that has a hole through its center and fins on its side to prevent turning during installation.
    c. Expansion anchors, commonly called molly screws (figure 4-25).
    d. Conical screws have deep threads that are screwed into gypsum board, strand board, and similar material (figure 4-26).
    e. Universal plugs (figure 4-27) are made of nylon and used for a number of hollow wall and some solid wall applications.

VIDEO #2
VIDEO #3

H. Connectors

1. Connectors are metal pieces formed into various shapes to join wood to wood, or wood to concrete or other masonry. They have specific names depending on their function.
2. Wood-to-Wood
    a. Framing anchors and seismic and hurricane ties (figure 4-28) are used to join parts of a wood frame.
    b. Post and column caps and bases are used at the top and bottom of those members (figure 4-29).
    c. Joist hangers and beam hangers are used to support joists and beams (figure 4-30).
3. Wood-to-Concrete
    a. Sill anchors, anchor bolts, and holdowns (figure 4-31) connect frame members to concrete.
    b. Girder hangers and a beam seat (figure 4-32) make beam-to-foundation wall connections.
    c. Post bases are used to anchor posts to concrete floors or footings.

I. Adhesives

Demonstrate common gluing process such as an edge-to-edge lamination. When cured, sand the joint to show how joint virtually disappears. Break the laminated pieces to show strength of glue joint.

1. Glue is used mostly for interior finish work. A number of mastics are used throughout the construction process.
2. Glue
   a. White and yellow glue are used most frequently by carpenters.
      - Fast setting.
      - For interior use.
      - Some varieties are moisture-resistant yet should not be used for exterior applications.

   b. Contact Cement
      - Bonds on contact.
      - Need not be clamped under pressure.
      - Widely used to apply plastic laminates for kitchen countertops.
3. Mastics

VIDEO #3
VIDEO #8

   a. Come in cans or cartridges.

   b. Construction Adhesive
      - Used in glued floor system.
      - Can be used in cold weather, even on wet or frozen wood.
      - Also used on stairs to increase stiffness and eliminate squeaks.

   c. Panel Adhesive (figure 4-33)
      - Used to apply such things as wall paneling, foam insulation, gypsum board, and hardboard to wood, metal, and masonry.

   d. Troweled Mastics
      - A notched trowel is used to spread the adhesive.
      - Depth and spacing of notches on trowel determine amount of adhesive applied.
      - Used for installing ceramic tile, vinyl base, and vinyl floor tile.

Chapter 5 Blueprints, Codes, and Building Layout

Instructor's Note

In this column, you will find a variety of lecture tips to aid student comprehension, notes on when to use PowerPoint slides to teach concepts, safety cautions for your students, and notes on when your class should view a video segment.

Appendix A lists corresponding numbered PowerPoint Presentation titles (see page 197)

Appendix B lists corresponding numbered Video titles and sections as applicable (see page 201)

Introduction

1. Carpenters must be able to picture what the building will look like in order to build to specification.
2. The term blueprint comes from an early processing of paper that turned the entire sheet blue except for where lines and words were drawn.
3. Building codes and ordinances were developed to ensure that buildings are built according to safety and design standards for a particular locale.
4. The first step in construction is locating where the building will be laid on the lot; this is done by shooting elevations.

OBJECTIVES

- Describe and explain the function of the various kinds of drawings contained in a set of blueprints.
- Demonstrate how specifications are used.
- Identify various types of lines and read dimensions.
- Establish level points across a building area using a water level and by using a carpenter's hand spirit level in combination with a straightedge.
- Accurately set up and use the builder's level, transit-level, and laser level.
- Use an optical level to determine elevations.
- Lay out building lines by using the Pythagorean Theorem and check the layout for accuracy.
- Build batter boards and accurately establish building lines with string.
- Identify and explain the meaning of symbols and abbreviations used on a set of prints.
- Read and interpret plot, foundation, floor, and framing plans.
- Define and explain the purpose of building codes and zoning laws.
- Explain the requirements for obtaining a building permit and the duties of a building inspector.

Glossary of Blueprints, Codes, and Building Layout

Detail – close up view of a plan or section

Elevation – a drawing in which the height of the structure or object is shown; also, the height of a specific point in relation to another point

Foundation – that part of a wall on which the major portion of the structure is erected

Laser – a concentrated, narrow beam of light; laser-equipped devices are used in building construction

Ledger – a temporary or permanent supporting member for joists or other members running at right angles; horizontal member of a set of batter boards

Plan – in an architectural drawing, an object drawn as viewed from above

Pythagorean Theorem – a mathematical expression that states that the sum of the square of the two sides of a right triangle equals the square of the hypotenuse

Section – drawing showing a vertical cut through an object or part of an object

A. Blueprints

Obtain an actual set of blueprints and review with students.

1. Different kinds of drawings are required for a construction of a building. These are called plans, elevations, sections, and details. When put together, they constitute a set of prints.
2. Plans
   a. Plans are views from a structure as viewed from above.
   b. The plot plan (figure 5-1) shows information about the lot, e.g., location of the building, walks, and driveways.
   c. The floor plan shows a horizontal cut of the structure four to five feet above the floor (figure 5-2).
   d. The foundation plan shows a horizontal cut of the foundation walls (figure 5-3).
   e. The framing plan shows the direction and spacing of floor and roof framing members (figure 5-4). They are not always included.
3. Elevations
   a. Elevations are drawings that show the shape and finishes of all sides of the exterior of the building (figure 5-5).
   b. Interior elevations are sometimes included that would usually show kitchen and bathroom elevations (figure 5-6).
4. Sections
   a. A section is a drawing that shows a vertical cut through all or part of the construction (figure 5-7).
5. Details
   a. Details are specific parts drawn at a larger scale to show a portion of construction more clearly (figure 5-8).
6. Other Drawings
   a. Drawings related to electrical work, plumbing, heating, and ventilating may be on separate sheets on a set of prints.
7. Schedules
   a. Window and door schedules give information about location, size, and kind of windows and doors to be installed throughout the building (figure 5-9).
   b. Finish schedule provides information on the kind of finish material to be used on the floors, walls, and ceilings of the individual rooms (figure 5-10).
8. Specifications
   a. Commonly called specs.
   b. Supplement the working drawings with more complete descriptions of methods, materials, and quality of construction.
   c. If there is a conflict of information, the specifications take precedence over the drawings.

Using graph paper, have students draw a simple floor plan to scale at your specifications.

9. Blueprint Language
   a. Scales
      - Most commons scale found on blueprints is 1/4 inch equals 1 foot.
      - Plot plan may be 1/16 inch equals 1 foot.
      - Section views and details may be 1/2 inch equals 1 foot, up to 3 inches equals a foot (figure 5-11).
   b. Types of Lines
      - Variation in width of line is called line contrast (figure 5-12).
      - Object lines are broad, solid lines that outline the object being viewed.
      - Hidden lines consist of short, fine, uniform dashes that indicate an object not visible.
      - Centerlines are indicated by a fine, long dash, then a short dash, then a long dash, and show centers of windows, doors, and partitions.
      - A section reference or cutting plane line is a broad line consisting of a long dash followed by two short dashes. Shows a cross-sectional view of a specific part of a building (figure 5-13).
      - A break line is used to terminate part of an object. Its purpose is to shorten the drawing to make maximum use of space.
      - A dimension line is a fine, solid line used to indicate the location, length, width, or thickness of an object (figure 5-14).
      - A leader line is a fine, solid line and points to an object from a location.
   c. Dimensions
      - Given in feet and inches.
      - Dimensions under one foot are given in inches (figure 5-15).
   d. Symbols
      - Used to represent objects in the building (figure 5-16).
   e. Abbreviations
      - Used to conserve space.
      - Location of the abbreviation is the key to its meaning (figure 5-17).

B. Codes

1. Cities and towns often have local laws governing many aspects of new construction and remodeling. There are state and national building codes. They are designed to protect the consumer and the community.
2. Zoning Regulations
   a. Keep buildings of similar size and purpose in areas for which they have been planned.
   b. Regulate space in each of the areas.
   c. Are shown on zoning maps.
   d. Zones
      - Vary from community to community.
      - Single family residential zones.
      - Multi-family residential zones.
      - Mobile home parks.
      - Combination of residences, retail stores, and offices.
      - Central business district.
      - Commercial districts.
      - Industrial zones.

   e. Lots
      - Most cities specify a minimum lot size for each zone and a maximum ground coverage for the structure.
      - Maximum height of the building may be stipulated.
      - Minimum lot width is usually specified, as are setbacks (distance from lot line).
      - Some communities require a certain amount of landscaped area called green space.
   f. Nonconforming Buildings
      - Because some cities were in existence before the advent of zoning laws, many buildings and businesses may not be in their proper zone. They are called nonconforming.
      - Any hardships imposed by zoning regulations may be relieved by a variance. Zoning Board of Appeals grants variances.

Using a book of codes, have students look up minimum stair width and headroom specifications for a residential building. Look up others at your discretion.

3. Building codes
   a. Regulate the design and construction of buildings by establishing minimum safety standards.
   b. Construction superintendents and contractors must have extensive knowledge of codes.
   c. National Building Codes
      - International Building Code (IBC) has been adopted in an attempt to standardize minimum building standards.
   d. Use of Residential Codes
      - In addition to structural elements, residential codes may include the following:
         1) Exits
         2) Emergency exits
         3) Smoke detectors
         4) Room dimensions
         5) Light
         6) Ventilation
         7) Sanitation
         8) Window size and placement
         9) Maximum limits of glass area
         10) Fans vented to the outside
         11) Requirements for baths and kitchens
         12) Requirements for hot and cold water
4. Building Permits

VIDEO #1
Section 1

   a. Building permit is needed before construction can begin.
   b. Requires a building permit application form (figure 5-18).
   c. Permit card must be displayed at the site in a conspicuous place until the construction is completed (figure 5-19).
   d. The fee for a permit is based on estimated cost of construction.
5. Inspections
   a. Building inspectors visit the job site to perform code inspections at various intervals.
      - Foundation inspection.

- Frame inspection.
- Final inspection.

b. Certificate of Occupancy or Completion is then granted.

C. Building Layout

Have students practice using builder's level, water level, and laser levels.

1. Leveling Tools
   a. Levels and straightedges.
      - If no other tools are available, a carpenter's hand level and a long straightedge may be used (figures 5-20 and 5-21).

VIDEO #1 Section 7

   b. Water level (figure 5-23).
   c. Builder's level (figure 5-24).
   d. Transit-level (figure 5-25).
   e. Automatic level (figure 5-26).

VIDEO #1 Section 4

2. Setting up and Adjusting the Optical Level
   a. Open and adjust the legs of the tripod to a convenient height.
   b. Secure instrument to tripod.
   c. Turn leveling screws to center the bubble in the spirit level (figure 5-27).
   d. Caution: Care must be taken not to damage the instrument. Never force any parts of the instrument.
   e. Caution: Make sure the feet of the tripod do not slip on smooth or hard surfaces.
   f. Caution: Do not leave a set-up instrument unattended around moving equipment.
   g. Sighting the Level:
      - Rotate the telescope and sight over its top, aiming at the object.
      - Look through telescope, focusing it by turning the knob.
      - Center the crosshairs of an object by moving the telescope left or right (figure 5-29).
      - Caution: If lenses need cleaning, dust them with a soft brush or rag. Do not rub the dirt off, as rubbing may scratch the lens coating.
3. Using the Optical Level
   a. When the instrument is leveled, a given point on the line of sight is exactly level with any other point (figure 5-30).
   b. Targets
      - Stick (figure 5-31).
      - Folding rule.
      - Steel tape.
      - Leveling rod.

VIDEO #1 Section 6

   c. Establishing Elevations
      - Done by establishing a benchmark.
   d. Laying out a Horizontal Angle
      - Called a vertex.
      - Establish one side of the angle.
        1) Tighten clamp screw.
        2) Turn horizontal scale to zero.
        3) Loosen clamp screw.
        4) Swing telescope 90 degrees.

5) Tighten clamp screw.
6) Establish opposite side of the angle (figure 5-32).

VIDEO #1
Section 8

e. Laser Level
- A laser level is a device that releases a narrow beam of light.
- Developed to provide more accurate and efficient layout work (figure 5-33).
- Kinds of laser levels
  1) Low price models are adjusted manually.
  2) More expensive models automatically adjust and maintain level.
- Establishing and determining elevations
  1) Indoors, a red beam will create a level plane of sight.
  2) Outdoors, an electric sensor, also called a receiver or detector, is attached to the leveling rod or stick (figure 5-35).
- Laser safety (figure 5-36)
  1) Never stare directly into the laser beam or view it with optical instruments.
  2) When possible, set the laser up so it is above or below eye level.
  3) Turn the laser off when not in use.
  4) Do not point the laser at others.

VIDEO #1
Section 2

D. Locating the Building

SLIDE 5

1. Layout begins with the dimensions of the building and its location on the site.
2. This information is determined from the plot plan. Often it is the carpenter's responsibility to lay out building lines.

VIDEO #1

3. Staking the Building

Stake out a building and set up batter boards for that building.

a. Find survey markers that locate the corners of the property. Measure in on each side from the front property line to the specific front setback (figure 5-37).
b. Drive stakes on each end. Stretch a line between the stakes to represent the front line of the buildings (figure 5-38).

SLIDE 8

c. Along the front building line, measure in from the side property line the specified side setback distance of the building. Drive the first building corner stake (figure 5-39).
d. From the first stake, measure the dimension of the building length along the front building line. Drive the second stake (figure 5-40).

VIDEO #1
Section 4

e. The third stake on the rear building line may be located by using one of three methods:
- Transit-level (figure 5-32).
- 3—4—5 method (figure 5-41).
- Pythagorean Theorem (figure 5-42).

VIDEO #6
Section 5

f. Location of the fourth stake is done using two tapes. Measure the building length from stake 3 and the building width from stake 2 (figure 5-43).

SLIDE 9

g. Check squareness of building with diagonal measurements. Make adjustments if the measurements differ while maintaining accuracy of width and length measurements (figure 5-44).
h. All measurements must be made on the level. If the land slopes, the tape is held level with a plumb bob suspended from it (figure 5-45).
- Irregularly-shaped buildings are laid out using the same fundamental principles (figure 5-46).

VIDEO #1
Section 6

4. Batter Boards
   a. Installed to allow layout stakes to be reinstalled quickly after excavation is made.
   b. Consist of horizontal members called ledgers. These are attached to stakes driven in the ground (figure 5-47).
   c. Set up the builder's level about center on the building location.
   d. Site to benchmark and determine height of ledgers.
   e. Install ledgers level on stakes.

VIDEO #1
Section 9

   f. Stretch lines between batter boards directly plumb over original corner stakes.
   g. Check accuracy of layout, adjust if necessary (figure 5-49).
   h. Make saw cut on outside corner of the top edge of the ledger.

Chapter 6 Concrete Form Construction

Instructor's Note

In this column, you will find a variety of lecture tips to aid student comprehension, notes on when to use PowerPoint slides to teach concepts, safety cautions for your students, and notes on when your class should view a video segment.

Appendix A lists corresponding numbered PowerPoint Presentation titles (see page 197)

Appendix B lists corresponding numbered Video titles and sections as applicable (see page 201)

Introduction

1. Concrete as a building material is unique.
2. It must be placed right the first time.
3. If a form fails to hold, costly labor time and materials are lost.
4. Form construction should be strong yet easily removed.

OBJECTIVES

- Construct forms for footings, slabs, walks, and driveways.
- Construct concrete forms for foundation walls.
- Lay out and build concrete forms for stairs.
- Explain techniques used for the proper placement and curing of concrete.
- Describe the composition of concrete and factors affecting its strength, durability, and workability.
- Explain the reasons for making a slump test.
- Explain the reasons for reinforcing concrete and describe the materials used.
- Estimate quantities of concrete.

Glossary of Concrete Form Construction Terms

Buck – a rough frame used to form openings in concrete walls

Concrete – a building material made from portland cement, aggregates, and water

Concrete block – a concrete masonry unit (CMU) used to make building foundations typically measuring 8″ × 8″ × 16″

Frost line – the depth to which the ground typically freezes in a particular area; footings must be placed below this depth

Girder – heavy timber or beam used to support vertical loads

Gusset – a pad of wood or metal used over a joint to stiffen and strengthen it

Pilaster – column built within and usually projecting from a wall to reinforce the wall

Portland cement – a fine gray powder that when mixed with water forms a paste that sets rock hard; an ingredient in concrete

Reinforcing rods – also called rebar; steel bars placed in concrete to increase tensile strength

Rise – in stairs, the vertical distance of the flight; in roofs, the vertical distance from plate to ridge; may also be the vertical distance through which anything rises

Run – the horizontal distance over which rafters, stairs, and other like members travel

Scab – a length of lumber applied over a joint to stiffen and strengthen it

Spreader – a strip of wood used to keep other pieces a desired distance apart

Stud – vertical framing member in a wall running between plates

Vapor retarder – also called a vapor barrier, a material used to prevent the passage of moisture

VIDEO #2
Section 9

A. The Nature of Concrete

1. Concrete can be formed into practically any shape for construction of buildings, bridges, dams, and roads.

Obtain MSDS (Material Safety Data Sheets) on concrete.

2. Concrete is a mixture of portland cement, fine and coarse aggregates, water, and various admixtures.
3. Most concrete used in construction is ready-mixed concrete delivered to the job in a large truck (figure 6-2). It is sold by the cubic yard.
4. Concrete should be placed on a stable subsoil.

B. Types of Concrete Forms

1. Footing Forms

Have students build a simple rectangular form for a column footing. Fill with wet gravel to test strength.

   a. Footing provides a base on which to spread the load of the structure over a wider area of the soil.
      - For foundation walls, a continuous or spread footing is used (figure 6-3).
      - For columns and posts, pier footings are used.
      - The footing width is usually twice the wall thickness. The footing depth is usually equal to the wall thickness (figure 6-4).
      - Footings are strengthened by reinforcing rods of specified size and spacing.

   b. Footings must be located below the frost line (figure 6-5).

   c. Keyways
      - Usually formed in the footing by pressing a 2 × 4 into the fresh concrete (figure 6-6).
      - A keyway is to provide a lock between the footing and the foundation wall.

   d. Stepped Wall Footings
      - When a foundation is to be built on sloped land, it is sometimes necessary to step the footing.
      - The vertical and horizontal footing distances are adjusted so that a whole number of concrete blocks or concrete forms can be placed into that section of the footing (figure 6-7).

Have students build a small pier form or tapered pier form. Fill with wet gravel to test for strength.

   e. Column Footings
      - Forms supporting columns, posts, fireplaces, or chimneys are usually built by nailing form pieces together in square, rectangular, or tapered shapes to the specified size (figure 6-8). Stakes are driven on all sides.

2. Slab Forms

Slab on grade: Have students form sidewalk or monolithic slab.

   a. Slab-on-Grade: Where frost penetration into the ground is not very deep, a concrete slab may be placed directly on the ground (figure 6-9).

VIDEO #2
Section 1

   b. Monolithic Slab
      - A combined slab and foundation/footing is called a monolithic slab (figure 6-10).
      - It is also referred to as a thickened edge slab.

VIDEO #2
Section 4

   c. Slab Protection
      - Concrete exposed to living space of a building should be protected from moisture and cold.
      - Extruded polystyrene is the only insulation recommended for ground contact. It is placed between concrete and the subsoil.

Have examples of extruded polystyrene and vapor barrier such as 6 mil poly.

VIDEO #2
Section 4

- A vapor retarder (also called a vapor barrier) is placed under the concrete slab to prevent moisture from rising through the slab. It also provides a barrier to soil gases such as radon and methane entering the living space.
- The slab should be reinforced with mesh or rebar.
- The concrete must be at least 4 inches thick and haunched under load-bearing walls (figure 6-11).

3. Wall Forms
   a. Wall forms are manufactured of wood, aluminum, or steel.
   b. Specially designed hardware is used for spacing, forming, aligning, and bracing the panels.
   c. Wall Form Components

Have examples of panels, snap ties, and walers to show students.

- Form Panels
  1) Panels built of special form plywood, backed by metal ribs or 2 × 4 studs.
  2) They are placed side-by-side to form the inside and the outside of the foundation walls.
- Snap Ties
  1) Snap Ties hold the wall forms together at a desired distance apart (figures 6-12 and 6-13).
- Walers
  1) Walers are doubled 2 × 4 pieces with spaces in between them used to stiffen the forms (figure 6-14).

4. Concrete Forming Systems
   a. Consists of manufactured panels and components for concrete form construction.
   b. Panels may be made of steel or a combination of steel and wood, and are tied together with metal wedges (figure 6-15).

Have an example of manufactured panel system if available.

   c. Starting at the outside corner, panels are wedged together and act as yokes (figure 6-16).
   d. Walers are installed (figure 6-17).
   e. Adjustable braces are installed to hold the wall straight (figure 6-18).
   f. Steel inside corners are installed (figure 6-19).
5. Stair Forms
   a. Concrete stairs may be suspended or supported by earth (figure 6-20).
   b. Rise is the height of each step; it is called the riser.
   c. Run is the horizontal distance of the step; it is called the tread.
   d. Layout may be done with a chalk line (figure 6-21).

Determine rise, run, riser height, and tread length of various stairs.

C. Economy and Conservation in Form Building

1. Economical concrete construction depends on the reuse of forms.
2. Care must be taken when stripping form panels to prevent damage.
3. Strip forms should be clean of all adhering concrete and stacked neatly.

VIDEO #2
Section 11

D. Concrete Reinforcement

Demonstrate and/or have students identify, cut, and bend a sample piece of rebar.

1. Concrete has a high compressive strength and a low tensile strength.
2. Rebars
   a. Steel reinforcing bars are called rebars.
   b. Rebars are used in concrete to increase tensile strength.
   c. Concrete is then called reinforced concrete.

VIDEO #2
Section 5

VIDEO #2
Section 5

d. Rebars come in standard sizes, identified by numbers that indicate the diameter in eighths of an inch. Example: Number 6 rebar has a diameter of 6/8 or 3/4 inch (figure 6-23).

e. Size, location, and spacing is shown on the plan.

f. Placed inside the form before concrete is placed.

Have an example of welded wire mesh to show students.

3. Wire Mesh

a. Used to reinforce concrete floor slabs resting on the ground.

b. Is laid in the slab above the vapor retarder before the concrete is placed.

Mix a small amount of concrete and perform a slump test using a slump cone.

E. Placing Concrete

1. Inside surfaces of the forms are sprayed with form oil to make form removal easier. Reinforcement should be free of oil.

2. Slump tests are made by inspectors on the job to determine the consistency of the concrete (figure 6-25).

3. Rate of Placement

VIDEO #2
Section 11

a. Concrete must not be placed at a rapid rate, especially in high forms.

b. The amount of pressure at any point on the form is determined by the height and weight of the concrete above it (figure 6-26).

c. To eliminate voids or honeycombs in the concrete, it should be worked by hand-spading or vibrated after it goes into the form.

d. Overvibrating increases the lateral pressure on the forms (figure 6-28).

4. Curing Concrete

a. Concrete is cured either by keeping it moist or by preventing loss of its moisture for a period of time.

b. Flooding or constant sprinkling of the surface is the most effective method of curing concrete.

c. Concrete surfaces may be covered with plastic film or other waterproof sheets.

d. Liquid curing chemicals may be used to seal in moisture and prevent evaporation. Caution: Curing agents placed on green concrete have chemicals that are harmful. It is important to follow the manufacturer's directions.

e. Protect concrete from freezing for at least four days after being placed.

VIDEO #2
Section 10

5. Precautions Using Concrete

a. Wear protective clothing when working with newly mixed concrete.

b. Wash skin areas that have been exposed to concrete as soon as possible.

VIDEO #2
Section 8

c. If any material containing cement gets into the eyes, flush immediately with water and get medical help.

F. Foundation Types

1. Concrete Block Foundation

a. Made by stacking together concrete masonry units (CMU) in an array (figure 6-29).

b. CMUs measure 8 inches high and typically 16 inches long; some are 18 inches long.

c. Each measurement is smaller to allow for a 3/8-inch mortar joint.

d. CMU widths range from 4 to 12 inches in 2-inch increments (figure 6-30).

e. Most CMUs are called stretchers, which have the familiar shape of a CMU. Some are constructed as sash or corner blocks (figure 6-31).

2. All Weather-Wood Foundation

a. Made of pressure-treated lumber and plywood (figure 6-32).

b. Framing techniques are the same as for walls and floors (see Chapters 7 and 8).

3. Precast Concrete Foundations
    a. Manufactured off-site and delivered on truck, then craned into place by a specially trained crew (figure 6-33).
    b. Sections are insulated with 1-inch of extruded polystyrene foam.
    c. Each section is bolted together and each seam is caulked to produce a guaranteed watertight seal (figure 6-34).
    d. A slab is later placed as a floor to support the bottom of the foundation from moving towards the inside during the backfilling of the foundation.
4. Insulated Concrete Forms
    a. Uses 2-inch thick expanded polystyrene (figure 6-35) manufactured into blocks with a metal or plastic tie system.
    b. Block sizes vary depending on manufacturer ranging from whatever wall thickness needed × 16–18″ high × 3–4 feet long (figure 6-36).
    c. Fastening strips included on their surfaces make it easy to screw on a wall finish later.
    d. Rebar is installed as needed after each block layer is installed.
    e. Placement of the concrete is done in levels or lifts.
    f. Only one course is filled at a time, giving the concrete time to hydrate or harden slightly.

SLIDE 2

G. Estimating Concrete

1. Ready-mix concrete is sold by the cubic yard or cubic meter.
2. To determine the number of cubic yards of concrete needed for a job, find the number of cubic feet and divide by 27.

Determine the number of cubic yards for continuous footing, slab-on-grade, monolithic slab, wall form, and tapered pier form.

Have students construct a footing form.

H. Procedures

1. Constructing Footing Forms
    a. Stretch lines back on the batter boards.
    b. Suspend a plumb bob from the batter boards at each corner (figure A).
    c. Drive stakes to the correct elevation and stretch lines between the stakes (figure B).
    d. Begin forming with the outside form board, using stakes to form the sides in position. Space stakes 4 to 6 feet apart.
    e. Cut a number of spreaders.
    f. Erect inside forms in a manner similar to outside forms. Fasten the spreaders across the forms at intervals necessary to hold the form a correct distance apart (figure C).
    g. Footing forms are braced with wood and by shoveling earth against the outside of the forms (figure D).

VIDEO #2
Section 1

VIDEO #2
Section 1

I. Erecting Wall Forms

Have students erect and brace wall forms.

1. After the footing concrete has hardened sufficiently, remove and clean the footing forms.
2. Stretch lines again on the batter boards.
3. Suspend the plumb bob from the layout lines to the footing.
4. Snap a chalk line on the top of the footing (figure A).
5. Set panels directly on the concrete footing to the chalk line or on 2 × 4 or 2 × 6 lumber plates (figure B).
6. Stack the panels necessary to form the inside of the walls and the outside of the walls. Caution: Form panels are heavy and care should be taken while lifting them.

7. Set all outside corner panels in place (figure C).
8. Fill in between the corners with panels, placing snap ties in the slots between panels as work progresses.
9. Tie rebars in place as detailed on prints.
10. Erect panels for the inside of the wall, inserting other end of snap ties as work progresses (figure D).
11. Install the walers (figure E).
12. Reinforce the corners with vertical 2 × 4s. This is called yoking the corners (figure F).
13. Straighten forms using a string (figure H).
14. Brace the walls inside and outside as necessary to straighten them (figure I).
15. Snap chalk lines on the inside of the form panels to the height of the foundation wall (figure J).
16. Place concrete in the recommended procedure and level to determined height.
17. Caution: Do not overvibrate the concrete. This will cause the aggregate to separate and flow to the bottom, or cause the form to rupture.
18. Set the anchor bolts into fresh concrete (figure K). Bolts should be spaced 6 to 8 feet apart, and 6 to 12 inches from each corner.
19. Install blockouts for each opening. Blockouts are also called bucks (figure L).
20. Install girder pocket blockouts (figure M).

J. Constructing Concrete Stair Forms

Have students lay out and construct a stair form.

1. Determine rise and run of each step and lay them out on the inside of the existing walls or form.
2. Secure cleats.
3. For earth supported stairs, place fill and compact (figure A).
4. For suspended stairs, construct a support made of plywood, joists, shores, and stair horses (figure B).
5. Brace all members into position.
6. Oil all wood and plywood to facilitate stripping.
7. Install the rebars or reinforcing as specified.
8. Install riser boards.
9. Brace the risers from top to bottom at mid-span (figure C).

Chapter 7 Floor Framing

Instructor's Note

In this column, you will find a variety of lecture tips to aid student comprehension, notes on when to use PowerPoint slides to teach concepts, safety cautions for your students, and notes on when your class should view a video segment.

Appendix A lists corresponding numbered PowerPoint Presentation titles (see page 197)

Appendix B lists corresponding numbered Video titles and sections as applicable (see page 201)

Introduction

1. Wood frame construction is widely used for residential and light commercial construction.
2. Because of the ease with which wood can be cut, shaped, fitted, and fastened, many different architectural styles are possible.
3. The floor frame is the first section installed after the foundation is completed. The floor system has many components.
4. In order to assemble a floor frame, a carpenter must be able to identify all the components, accurately locate their positions, and cut each member to fit.

OBJECTIVES

- Describe platform, balloon, and post-and-beam framing, and identify framing members of each.
- Describe several energy and material conservation framing methods.
- Build and install girders, erect columns, and lay out sills.
- Lay out and install floor joists.
- Frame openings in floors.
- Lay out, cut, and install bridging.
- Apply subflooring.
- Describe methods to prevent destruction by wood pests.

Glossary of Framing Terms

Anchor bolt – long metal fastener with a threaded end used to secure materials to concrete

Balloon frame – a type of frame in which the studs are continuous from foundation sill plate to roof

Band joist – the member used to attached to the ends of floor joists where they rest on the sill

Box header – the member attached to the ends of floor joists where they rest on the sill

Bridging – diagonal braces or solid wood blocks between floor joists used to distribute the load imposed on the floor

Column – a large vertical member used to support a beam or girder

Dimension lumber – a term used to describe wood that is sold for framing and general construction

Draftstops – (also called firestops) material used to reduce the size of framing cavities in order to slow the spread of fire; in a wood frame, might consist of full width dimension lumber blocking between studs

Flush – a term used to describe when surfaces or edges are aligned with each other

Girder – a heavy timber or beam used to support vertical loads

Header – pieces placed at right angles to joists, studs, and rafters to form and support openings in a wood frame

Joist – horizontal framing members used in a spaced pattern that provide support for the floor or ceiling system

Linear feet – a measurement of length

Masonry – any construction of stone, brick, tile, concrete, plaster, and similar materials

On center – the distance from the center of one structural member to the center of the next one; usually abbreviated as OC or o.c.

Plate – top or bottom horizontal member of a wall frame

Post – a vertical member used to support a beam or girder

Pressure-treated – treatment given to lumber that applies a wood preservative under pressure

Ribbon – a narrow board let into studs of a balloon frame to support floor joists

Sheathing – boards or sheet material that are fastened to joists, rafters, and studs and on which the finish material is applied

Shim – a thin, wedge-shaped piece of material used behind pieces for the purpose of straightening them, or for bringing their surfaces flush

Sill – first horizontal wood member resting on the foundation supporting the framework of a building; also, the lowest horizontal member in a window or door frame

Sill sealer – material placed between the foundation and the sill to prevent air leakage

Subfloor – material used as the first floor layer on top of joists

Tail joists – shortened on-center joists running from a header to a sill or girder

Termite shields – metal flashing plate over the foundation to protect wood members from termites

Trimmer – a joist or stud placed at the sides of an opening running parallel to the main framing members

A. Types of Frame Construction

SLIDE 10

1. Platform Frame Construction
   a. Platform frame is most commonly used frame in residential construction (figure 7-1).
   b. In this type of construction, the floor is built and then the walls are built on top of it.
   c. A disadvantage of the platform frame is a relatively large amount of settling caused by the shrinkage of the large number of horizontal load-bearing frame members.
   d. To reduce shrinkage, lumber with the proper moisture content should be used.
2. Balloon Frame Construction
   a. In the balloon frame, the wall studs and the first floor joists rest on the sill.
   b. The second floor joists rest on a 1 inch × 4 inch ribbon that is cut flush with the inside edges of the studs (figure 7-3).
   c. Draftstops (sometimes called firestops) must be installed in walls in several locations.
   d. Used to prevent the spread of fire, a draftstop is usually made of dimension lumber (figure 7-4).

3. Post-and-Beam Frame Construction
   a. The post-and-beam frame uses fewer but larger pieces that are widely spaced (figure 7-5).
   b. Floors are sheathed with APA rated Sturd-I-Floor, 48 inches on center, which is 1 3/32 inch thick (figure 7-6).
   c. Walls
      - Exterior walls may be constructed with widely-spaced posts.
      - Usually some sections between posts in the wall are studded, as in platform framing (figure 7-7).
      - This provides for door openings, fastening for finish, and wall sheathing (figure 7-7).
   d. Roofs
      - Post-and-beam framed roof is widely used.
      - Roof beams and sheathing are usually exposed.
      - Roof planks come in 2-inch, 3-inch, and 4-inch nominal thicknesses.
      - The roof is insulated on top of the roof deck (figure 7-8).
4. Energy and Material Conservation Framing Methods
   a. Conserve energy and use less material and labor.
   b. All framing, floors, walls, and roofs are spaced 24 inches OC (figure 7-10).
   c. Floors – Single layer of 3/4 inch tongue-and-grooved plywood is used over joists spaced 24 inches OC (figure 7-12).
   d. Walls – 24-inch OC spacing; wall opening laid out so at least one side of the opening falls on an on-center stud (figure 7-13).
   e. Roof – Rafters or trusses are spaced 24 inches OC.

Have engineered I-beam examples to show students.

B. Layout and Construction of the Floor Frame

1. A floor frame consists of members fastened together to support the loads a floor is expected to bear. Because platform framing is used more than any other type, this section describes how to lay out and construct its floor frame. The knowledge gained in this section can be used to lay out and construct any type of floor frame.

VIDEO #3
Section 3

2. Description and Installation of Floor Members (figure 7-15)
   a. Girders
      - Girders are heavy beams that support the inner ends of the floor joists.
      - May be made of solid wood, built up of 2 or more pieces of dimension lumber, laminated veneer lumber or Glulam beams, or I-shaped steel beams (figure 7-16).
      - A pocket formed in the foundation wall usually supports the ends of the girder (figure 7-18). The pocket should provide at least a 4-inch bearing for the girder and 1/2 inch clearance on both sides and the end.

VIDEO #3
Section 1

   b. Sills
      - Also called mudsills or sill plates.
      - They lie directly on the foundation wall.
      - Decay-resistant material such as pressure-treated lumber, redwood, black locust, or cedar.
      - Usually consists of single or doubled 2 × 6s (figure 7-19).
      - Attached to the foundation with anchor bolts.
      - Sill sealer installed between sill and foundation.

VIDEO #3
Section 4

3. Floor Joists
   a. Horizontal members of a frame.

VIDEO #3
Section 4

b. Two-inch nominal thick lumber most common; however wood I-beams are gaining popularity (figure 7-20).

c. Generally spaced 16 inches OC. May also be placed 24 inches OC.

d. Ends of floor joists are capped with a band joist, also called a rim joist, box header, or joist header (figure 7-21).

e. If joists are lapped over the girder, the minimal amount of overlap is 4 inches and the maximum overhang is 12 inches.

f. To gain more headroom, joists may be framed into the side of the girder (figure 7-23).

g. Notching and Boring of Joists
   - Notches in the bottom or top of sawn lumber floor joists should not exceed 1/6 of the joist depth.
   - Notches should not be located in the middle 1/3 of the joist span.
   - Notches on the ends should not exceed 1/4 of the joist depth.
   - Holes bored in joists should not be larger than 1/3 of the joist depth.
   - Holes bored in joists should not be closer than 2 inches to the top or bottom of the joists.
   - Wood I-beams are often manufactured with perforated knockouts in the web. To cut other size holes in the web, consult the manufacturer's specifications guide.

h. Doubling Floor Joists
   - For added strength at well openings and under partitions.
   - The top edges must be flush or even.
   - To bring them flush, toenail down through the top edge of the higher one at about the joist's mid-span (figure 7-25).

i. Cantilevered Joists
   - When the floor system extends beyond the foundation, the floor joists become cantilevered joists.
   - Are supported at the inside end by a double joist or girder (figure 7-26).
   - The joist hanger should be turned upside down from the normal installation.

Provide various wood connector samples and explain their specific purpose (i.e., location).

4. Wood Connectors

   a. There are many types and styles of connectors used to secure structural components of a building (figure 7-27).

   b. The size, quantity, and location of nails are very important.

   c. Follow manufacturer's recommendations.

5. Bridging

   a. Is installed in rows between floor joists at intervals not exceeding 8 feet.

   b. May be solid wood, wood cross-bridging, or metal cross-bridging (figure 7-28).

6. Columns

   a. Girders may be supported by framed walls, wood posts, or steel columns (figure 7-29).

   b. Metal plates are used at the top and bottom of columns to distribute the load over a wider area.

VIDEO #3
Section 5

7. Subflooring

   a. Is used over joists.

   b. Tongue-and-groove-edged Sturd-I-Floor panels are most commonly used.

8. Estimating materials

   a. Floor joists:

- First divide the length of the building by the spacing and add one.
- Next multiply the number of rows of floor joists.
- Last, add the number needed for doubling and for band joists.

b. Bridging – multiply the length of the building by 3 for each row of bridging.

c. Panel Subfloor – divide the area of the floor by 32.

C. Termites

1. Of all the destructive wood pests, termites are the most common. They are more prevalent in the warmer sections of the country (figure 7-30). All the techniques used for the prevention of termite attack are based on keeping the wood in the structure dry (equilibrium moisture content) and making it as difficult as possible for termites to get to the wood (figure 7-31).
2. Techniques to Prevent Termites

a. The Site

- All trees, roots, branches, and wood debris should be removed from the building site (not buried).
- Site should be graded to slope away from the building.
- Perforated drain pipe should be placed around the foundation alongside the footing (figure 7-33).
- Gutters and downspouts should be installed.

b. Exterior Woodwork

- Do not extend wall siding more than 2 inches below the top of the foundation wall.
- Keep woodwork at least 6 inches above the finished grade.
- Wood parts should never extend into or through concrete floors.

c. Termite Shields

- Force termites into the open (figure 7-34).
- Check local building codes that may mandate their use.

d. Use of Pressure-Treated Lumber (figure 7-35)

- Termites will generally not eat pressure-treated lumber.
- Building codes require the use of pressure-treated lumber for the following building members:

1) Wood joists located closer than 18 inches to exposed soil.

2) Wood girders that are closer than 12 inches to exposed soil.

2) Sleepers, sills, and foundation plates on a concrete or masonry slab that is in direct contact with the soil.

VIDEO #3
Section 3

D. Procedures

1. Installing Girders

a. Caution: Girders are heavy. Care must be taken when lifting them. Also make certain any temporary posts and bracing used to support the girder are secure enough to keep them from falling.

b. Set one end in the pocket of the foundation wall.

c. Place and fasten the other end in a braced temporary support.

d. Continue building and erecting sections until the girder is completed to the opposite pocket (figure A).

e. Temporary posts should be strong enough to support the weight imposed on them until the permanent ones are installed (figure B).

Review MSDS sheets covering pressure-treated lumber.

SLIDE 11

VIDEO #3
Section 1

2. Installing Sills
   a. Sills must be installed so that they are straight, level, and to the specified dimensions of the building.
   b. Snap a chalk line on top of the foundation wall in line with the inside edge of the sill.
   c. Cut the sections to length.
   d. Caution: Follow these safety rules when handling pressure-treated lumber.
      - Wear eye protection and a dust mask when sawing or machining treated wood.
      - When the work is completed, wash areas of skin contact thoroughly before eating or drinking.
      - Clothing that accumulates sawdust should be laundered separately from other clothing and before reuse.
      - Dispose of treated wood by ordinary trash collection or burial.
      - Do not burn treated wood.
   e. Hold the sill in place against the anchor bolts.
   f. Square lines across the sill and measure the distance from the center of each bolt to the chalk line (figure A).
   g. Bore holes in the sill for each anchor bolt.
   h. Place the sill sections after installing the sill sealer.
   i. Replace the nuts and washers.
   j. Raise ends of girder so it is flush with the top of the sill plate (figure B).

If floor frame is not available, snap lines on shop floor to simulate sills and girders. Have students practice laying out and placing joists. Frame well openings to practice order of assembly.

SLIDE 12

VIDEO #3
Section 2

3. Laying out Floor Joists and Floor Openings
   a. Locations of the floor joists are often marked on the sill plate. Locate the openings and special framing before laying out the OC joists (figure A).
   b. Start the joist layout by measuring the joist spacing from the end of the sill (figure B).
   c. Measure back 1/2 thickness of the joist. Square a line across the sill (figure C).
   d. From the squared line, measure and mark the spacing of the joists along the length of the building (figure D).
   e. Identify the layout marks that are not for full-length floor joists.
   f. Check the mechanical drawings to make adjustments in the framing to allow for the installation of mechanical equipment (figure F).
   g. If joists are in line, Xs are made on the same side of the mark on both the girder and the sill plate on the opposite wall.
   h. If the joists are lapped, an X is placed on both sides of the mark at the girder and on the opposite side of the mark on the other wall (figure G).

SLIDE 13

4. Framing Floor Openings
   a. Openings in floors should be framed before the floor joists are installed.
   b. Fasten the inside trimmer joists in place.
   c. Mark the location of the header on the trimmers.
   d. Cut four headers (figure A).
   e. Place two headers, one for each end of the opening. Be sure that the first headers are installed farthest from the floor opening (figure B).
   f. Fasten the tail joists in position (figure C).
   g. Double up the headers (figure D).
   h. Finally, double up the trimmer joists.

VIDEO #3
Section 4

5. Installing Floor Joists
   a. Stack the necessary number of full-length joists at OC intervals along both walls (figure A).
   b. Joists are installed with the crown edge up.
   c. Keep the end of the floor joist back from the outside edge of the sill plate by the thickness of the band joist (usually 1 1/2 inches) (figure B).
   d. Toenail the joists to the sill and girder (figure C).
   e. When all floor joists are in position, they are straightened and tacked with 1 × 3s mid-span to the top of the joists (figure D).
6. Installing Band Joists
   a. Fasten the band joist into the end of each floor joist.
   b. Toenail band joist to sill plate (figure A).

Have students lay out and cut cross bridging out of 1 × 3 stock for various joist depths and spacings.

SLIDE 14

7. Laying out and Cutting Wood Cross-Bridging
   a. Determine the actual distance between floor joists and the actual depth of the joist (figure A).
   b. Hold the framing square on the edge of a piece of bridging stock.
   c. Mark the bridging stock along the tongue at a distance equal to the depth of the floor joist while holding the blade at the measurement equal to the distance between floor joists (figure B).
   d. Mark the bridging stock along the blade.
   e. Rotate the square and align the same dimension in the same fashion as before.
   f. Mark along the tongue (figure C).
   g. Make the actual length of the piece about 1/4 inch shorter than the layout.
   h. Bridging is cut using a power miter box.
8. Installing Bridging
   a. Solid Wood Bridging
      - Cut the pieces to length.
      - Install pieces in every other joist space on one side of the chalk line (figure A).
      - Install pieces in the remaining spaces on the opposite side of the line.
   b. Wood Cross-Bridging
      - Start two 6d nails in one end of the bridging.
      - Place it flush with the top of the joist on one side of the line and drive nails (nail only top end) (figure B).
      - Fasten other piece of bridging to other joist on opposite side of the line.
   c. Metal cross-bridging is fastened in a manner similar to that of wood cross-bridging.
9. Installing Columns
   a. Straighten the girder.
   b. Measure from column footing to bottom of girder.
   c. Deduct thickness of the top and bottom bearing places.
   d. Cut to length.
   e. Install columns in a plumb position under the girder and centered on the footing (figure B).
   f. Fasten the top bearing plate to the girder.

VIDEO #3
Section 5

10. Applying Plywood Subflooring
    a. Starting at the corner from which the floor joists were laid out, measure in 4 feet (figure A).
    b. Snap a line across the tops of the floor joists.
    c. Start with a full panel to the chalk line (figure B).
    d. Align the OC joists to their correct spacing before nailing.
    e. Start the second row with a 1/2 sheet to stagger the end joints.
    f. Minimum nail spacing is 6 inches along supported edges and 12 inches at intermediate supports.

Chapter 8 Wall and Ceiling Framing

Instructor's Note

In this column, you will find a variety of lecture tips to aid student comprehension, notes on when to use PowerPoint slides to teach concepts, safety cautions for your students, and notes on when your class should view a video segment.

Appendix A lists corresponding numbered PowerPoint Presentation titles (see page 197)

Appendix B lists corresponding numbered Video titles and sections as applicable (see page 201)

Introduction

1. During the wall-framing phase of construction, the building begins to take shape.
2. Exterior walls are constructed to the correct height, braced plumb, and straightened.
3. Window and door rough openings are framed to specified sizes.
4. The term rough work generally refers to framing, but does not imply that the work is crude.

OBJECTIVES

- Identify and describe the function of each part of the wall frame.
- Determine the length of exterior wall studs.
- Describe four different types of walls used in residential framing.
- Determine the rough opening width and height for windows and doors.
- Lay out the wall plates for partition intersections, openings, and OC studs.
- Describe several methods of framing corner and partition intersections.
- Assemble and construct a wall section.
- Erect and temporarily brace a wall section plumb and straight.
- Describe the function of and install blocking and backing.
- Apply wall sheathing.
- Lay out, cut, and install ceiling joists.
- Identify and describe the components of non-structural steel wall framing.
- Install a steel door buck.
- Estimate the materials needed for walls and ceiling framing.

Glossary of Wall and Ceiling Framing

Backing – strips or blocks of wood installed in walls or ceilings for the purpose of fastening or supporting trim or fixtures

Blocking – pieces of dimension lumber installed between joist and studs for the purposes of providing nailing surface for intersecting framing members

Gable end – the triangular-shaped section on the end of a building formed by the rafters in a common or gable roof and the top plate line

Gypsum board – a sheet product made by encasing gypsum in a heavy paper wrapping used to create the wall surface.

Joist hanger – metal stirrups used to support the ends of joists

Load-bearing – term used to describe a structural member that carries weight from another part of the building

Soffit – the underside trim member of a cornice or any such overhanging assembly

SLIDE 15

VIDEO #3
Section 7

A. Parts of a Wall Frame

1. Sometimes the names given to certain parts of a structure may differ according to the geographical area.
2. Plates
   a. The top and bottom horizontal members of a wall frame are called plates.
   b. The bottom member is called sole plate, bottom plate, or shoe.
   c. The top members are called top plates.
3. Studs
   a. Studs are vertical members of the wall frame.
   b. On Center studs that must be cut to allow for an opening are called cripple studs.
   c. Studs are usually 2 × 4 or 2 × 6.
   d. Studs are usually spaced 16 or 24 inches on center.

Show examples of laminated veneer lumber, parallel strand lumber, and Glulam beams.

4. Headers
   a. They form the top of rough openings.
   b. Headers must be strong enough to support the load above the opening (figure 8-2).
   c. Headers are made in many different ways depending on desired strength and available materials (figure 8-3).
5. Rough Sills
   a. Form the bottom of a window opening.
   b. They secure the top ends of the cripple studs.
   c. They carry little load.
6. Trimmers (Jacks)
   a. Trimmers or jacks are shortened studs that support the headers (figure 8-4).
7. Corner Posts (figure 8-6)
   a. Corner posts are built in a number of ways.
   b. They are constructed to allow for insulation to penetrate the corners.
8. Partition Intersections (figure 8-7)
   a. Wherever interior partitions meet an exterior wall.
   b. They provide for fastening of interior wall covering.
9. Ribbons (figure 8-8)
   a. Ribbons are horizontal members of the exterior wall frame in balloon construction.
   b. They are used to support the second floor joists.
   c. They are notched into the studs.
10. Corner Braces (figure 8-9)
    a. No wall bracing is required if rated panel wall sheathing is used.
    b. Types
       - Metal bracings, T- or L-shaped, require a saw kerf in the stud to allow them to lay flat when installed.
       - Let-in bracing is a continuous, diagonal 1 × 4 notched into the face of the studs, top plate, and sole plate.

Have students look up rough openings for various windows using a manufacturer-supplied handbook.

SLIDE 16

VIDEO #3
Section 12

B. Exterior Wall Framing

1. Plans or blueprints usually indicate the height from finished floor to finished ceiling. It is necessary to determine the length of the wall studs.
2. Determining the Size of Rough Openings
   a. A rough opening (RO) is an opening framed in a wall in which to install doors and windows.
   b. It is the carpenter's responsibility to determine the rough opening size for the particular unit from the information given in the door and window schedule.
   c. The rough opening size for exterior doors is typically 2 1/2 inches wider than the actual door width, and 2 1/2 inches taller than the door height (figure 8-11).
   d. Rough Opening Sizes for Windows (figure 8-12)
      - Many kinds of windows are manufactured by a number of firms.
      - It is best to consult the manufacturer's catalog to obtain the rough opening sizes.
      - The header height is typically the same as for the doors.
3. Determining Wall Type (figure 8-13)
   a. The carpenter must first determine if the wall to be laid out is to be load-bearing or non-load-bearing.
   b. Exterior walls are referred to as walls, and interior walls are referred to as partitions.
   c. Load-bearing walls are usually built first.
   d. Non-load-bearing walls are end walls that usually run parallel with the joists.
   e. Interior partitions are also load and non-load-bearing.
   f. Figure 8-15 notes the similarities and differences in laying out walls and partitions.

Using steel track and dry erase markers or chalk, have students practice laying out wall plates for various walls.

4. Laying out the Plates
   a. Layouts for floor joists, exterior walls, ceiling joists, and rafters should all start from the same corner of the building.
5. Assembling and Erecting Wall Sections
   a. The typical method of framing a wall is to precut the wall frame members, assemble the wall frame on the subfloor, and then stand the frame up.
   b. The frame is erected, and the corners are plumbed and temporarily braced.
   c. The walls are then straightened and braced between corners.
   d. Precutting Wall Frame Members
      - Studs may be purchased that are precut to length.
      - Determine lengths for studs, jacks, headers, cripples, and sills.
      - Cut parts and label according to window schedule.
      - Assemble corner posts, partition assemblies, and headers (figure 8-16).
   e. Assembling Wall Sections
      - Crown all studs.
      - Assemble rough openings.
      - Install corner posts and partition intersections.
      - Install studs.
      - Optional: Sheath or install permanent bracing.
   f. Erecting and Temporarily Bracing Wall Sections
      - Lifting jacks may be used.
      - Temporary diagonal braces usually remain in position until the exterior sheathing is applied.

SLIDE 17

VIDEO #3
Section 8

VIDEO #3
Section 10

VIDEO #3
Section 13

VIDEO #3
Section 13

6. Wall Sheathing
   a. Usually consists of 4 × 8 foot sheets of rated plywood panels.
   b. Sheathing panel may be applied horizontally or vertically.
   c. APA-rated plywood panels that serve as wall bracing must be nailed with a nail spacing of 6 inches apart on the edges and 12 inches apart on the intermediate studs (figure 8-17).
7. Shearwalls
   a. A shearwall is a heavily reinforced wall section designed to improve the lateral strength.
   b. They are sometimes used around garage door openings where the wall is narrow (figure 8-19).

VIDEO #6
Section 1

C. Interior Framing

1. Load-Bearing Partitions
   a. Support the inner ends of ceiling joists or wall joists.
   b. They are usually placed directly over the girder or the bearing partition in the lower level.
   c. These partitions are laid out and erected in a manner similar to exterior walls.
2. Non-Bearing Partitions
   a. Carry no load.
   b. Divide floor area into rooms.
   c. Openings may be framed with single member headers.

VIDEO #6
Section 7

3. Backing and Blocking
   a. Backing
      - Used to provide support of later fastening of trim or fixtures.
        1) Bathroom – faucets, shower heads, lavatories, water closets (figure 8-22).
        2) Kitchen – tops and bottoms of wall cabinets and for the tops of base cabinets (figure 8-23).
      - If the ceiling is to be built down to form a soffit.
   b. Blocking
      - Blocking is installed to provide supports for parts of the structure and for weathertightness and draft stopping.
      - Sometimes blocking serves as backing.
      - Blocking can be installed in the following locations:
        1) Solid bridging.
        2) Extra support for plywood panel edges that are not tongue-and-grooved (figure 8-25).
        3) Between ceiling joists to support the top ends of partitions that run parallel to and between joists (figure 8-26).
        4) Draftstop blocking between studs (figure 8-27).

D. Ceiling Joists

1. Ceiling joists generally run from the exterior walls to the bearing partition across the width of the building.
2. Methods of Installing Ceiling Joists
   a. Rafters are located over the studs, and the ceiling joists are fastened to the side of the rafters (figure 8-29).
   b. This binds the rafters and ceiling joists together into a rigid triangle and keeps the walls from spreading outward due to the weight of the roof (figure 8-28).

c. Typically the ceiling joists are half the building width and therefore must be joined over a beam or bearing partition.

d. Sometimes the ceiling joists are installed inline, and the joint must be scabbed (figure 8-30).

e. Most of the time, the ceiling joists lap over a bearing partition in the same manner as for floor joists.

f. This puts a stagger in the line in the ceiling joist and consequently in the rafters as well (figure 8-31).

g. Cutting the Ends of Ceiling Joists
   - The ends of ceiling joists on exterior walls usually project above the top of the rafter.
   - Ends must be cut to the slope of the roof, flush with or slightly below the top edge of the rafter.
   - Lay out the cut using a framing square. Cut one joist for a pattern.
   - Make sure the length of the taper cut does not exceed three times the depth of the member.
   - Also ensure that the end of the joist remaining after cutting is at least half the member's width (figure 8-33).

h. Framing Ceiling Joists to a Beam
   - If the beam is to project below the ceiling, it is installed in the same manner as a girder or load-bearing partition (figure 8-34).
   - If the beam is to be raised in order to make a flush and continuous ceiling below, the joists are butted into the beam and supported by a ledger strip or joist hanger (figure 8-35).

Review order of assembly for well openings and floor joists.

i. Openings
   - Openings in ceiling joists need to be made for such things as chimneys, attic access, disappearing stairs, or skylights.
   - Large openings are framed in the same manner as for floor joists.
   - For small openings, there is no need to double the joists or headings (figure 8-36).

j. Ribbands and Strongbacks
   - Ribbands are 2 × 4s installed flat on top of the ceiling joists.
   - They are placed at mid-span to stiffen the joists as well as to keep the spacing uniform.
   - With the addition of a 2 × 6 installed on edge, the ribband becomes a strongback (figure 8-37).

3. Layout and Spacing of Ceiling Joists

   a. Ceiling joists are installed alongside the rafters and securely fastened to them.

   b. This intersection should rest over the top of OC studs of exterior walls and load-bearing partitions.

   c. Layout should start from the same corner of the building as the floor joists and wall studs.

   d. Installing Ceiling Joists
      - Installed in a manner similar to floor joists.
      - First and last ceiling joists on each end of the building are placed inside the wall and installed in-line (figure 8-38).

E. Steel Framing

VIDEO #6
Section 7

1. Carpenters often frame interior non-load-bearing partitions and apply furring channels of steel. Different kinds of fasteners are used and some special types of tools are helpful.

VIDEO #6
Section 7

2. Steel Framing Components
    a. Studs (figure 8-39)
        - Manufactured in 25-, 22-, and 20-gauge.
        - The stud web has punchouts at intervals through which to run pipes and conduit.
        - Come in widths of 3 1/2 inches, 5 1/2 inches, 8 inches, 10 inches, and 12 inches, with 1/4-inch and 1 5/8-inch leg thicknesses.
        - Available in lengths in 8, 9, 10, 12, and 16 feet.
    b. Track
        - Top and bottom plates of a steel-framed wall are called runners or track.
        - Available in standard lengths of 10 feet.
    c. Channels (figure 8-40)
        - Used for lateral bracing of walls, the channel is inserted through the stud punchouts.
    d. Furring Channels (figure 8-41)
        - Furring channels are applied to walls and ceilings for the screw attachment of gypsum panels.
3. Framing Steel Partitions
    a. Lay out steel-frame partitions as you would wood-framed partitions.
    b. Cut with tin snips or a chop saw.
    c. Track is usually fastened to concrete with powder-driven fasteners.
    d. Place all full-length studs in position between track with the open side facing in the same direction and web punchouts aligned vertically (figure 8-42).
    e. Method of Framing Door Openings
        - Method of framing door openings depends on the type of door frame used, i.e., one-piece frame or a three-piece knocked-down frame (figure 8-43).
4. Metal Furring
    a. May be used on ceilings and walls by applying it at right angles to joists or studs.
    b. May be applied vertically or horizontally to masonry walls.

F. Structural Insulated Panels (SIPs)

1. Consist of two outer skins and an inner core of an insulating material (figure 8-46).
2. Produced in thicknesses from 4 1/2 inches to 12 1/4 inches. Panel sizes range from 4 feet by 8 feet up to 9 feet by 24 feet.
3. Prefabricated at the manufacturing plant with all windows and doors precut (figure 8-47).
4. Large panels require equipment to set them into position from the truck (figure 8-48).
5. Fire performance of SIPs is similar to other wood frame structures.
6. Easy to achieve an airtight home with SIPs (figure 8-49).
7. SIPs use wood more efficiently than sawn lumber.

G. Methods of Estimating Materials for Walls and Ceiling Joists

Estimate materials for wall and ceiling joists for a given house frame.

1. Studs
    a. Determine the total linear feet of exterior wall.
    b. If spaced 16" OC, estimate one stud for every linear foot of wall.
    c. For 24-inch spacing, divide the total linear feet of wall by 1.333.

2. Plates
   a. Multiply the total linear feet by 3.
   b. Add 5% for waste.
3. Headers and Sills
   a. Calculate the width for each opening and add 6 inches to each.
4. Wall Sheathing
   a. Multiply the total linear feet of wall by the wall height to find square feet of wall.
   b. Add it to the gable area using the triangle area formula (A = 1/2 × base × height).
   c. Deduct the area of any larger openings.
   d. Add 5% for waste.
   e. Finally, divide the total area by 32 to determine number of sheets.
5. Ceiling Joists
   a. Divide the length of the building by the joist spacing in feet.
   b. Add one to start.
   c. Double this number for the other half of the building if applicable.
6. Studs
   a. Determine the total linear feet of exterior wall.

VIDEO #3
Section 7

Have students practice determining stud, jack, header, sill, and cripple lengths for various walls.

H. Procedures

1. Determining Length of Studs for Platform Frame Construction (figure A)
   a. Begin with specified finish floor to finish ceiling height as specified in drawings.
   b. Add the thickness of the ceiling below the ceiling joists and the thickness of furring strips in ceilings if used.
   c. Add the thickness of the finish floor.
   d. Deduct the total thickness of the top plates and the sole plate to find the length of the stud.
2. Determining the Length of Studs for Balloon Frame Construction (figure B)
   a. Determine specified height from finish floor to finish ceiling in all the stories.
   b. Add the thickness of both finish ceilings.
   c. Add the thickness of both finish floors and subfloors.
   d. Add the width of all floor joists.
   e. Deduct the total thickness of the top plates.

VIDEO #3
Section 12

3. Determining the Size of Rough Openings (figure A)
   a. Windows
      - Consult window manufacturer's literature.
   b. Doors
      - Add 2 1/2 inches to the width of the door unit and 2 1/2 inches to the height of the door unit.

VIDEO #3
Section 9

Using steel track and dry erase markers or chalk, have students practice laying out wall plates for various walls.

4. Wall Plate Layout
   a. Snap lines on the subfloor (figure A).
   b. Cut, place, and tack two plates on the deck aligned with the chalk line (figure B).
   c. Make sure butt seams are centered on the stud (figure C).
   d. Determine center line dimensions of all openings in the wall (figure D).
   e. Measure and mark rough opening width.

   f. Mark O on the side of each rough opening line away from the opening center (figure E).
   g. Mark Xs on the side of the line away from center to show full-length studs on each side of openings (figure F).
   h. Mark for all intersecting partitions (figure G).
   i. Mark on center stud locations (figure H). Where openings occur, mark a C to indicate location of cripple studs (figure I).

SLIDE 18

5. Assembling Wall Sections
   a. Turn plates up on edge and separate them a distance equal to the stud length (figure A).
   b. Frame openings first to make assembly easier.

VIDEO #3
Section 10

      - Nail through the plate into the header.
      - Fasten the rough sills in position to the jack studs (figure B).
      - Fasten a king stud to each jack stud (figure C).
      - Nail cripple studs as needed between plates and sills (figure D).
   c. Install partition studs.
   d. Install remaining studs (figure E).
   e. Square wall section by checking both diagonals (figure F).
   f. Install let-in bracing if required (figure G).
6. Erecting and Bracing Wall Sections
   a. Remove toenails from top of plate.
   b. Lift wall section into place.
   c. Plumb and temporarily brace (figure A).
   d. Ensure that the sole plate is still on the chalk line.
   e. Nail sole plate to band or floor joists.
   f. Fasten end studs at corners.
   g. String and straighten walls (figure B).
7. Installing Backing and Blocking
   a. Snap a line across the framing.
   b. Blocking may be installed in a staggered row.
   c. Blocking may be installed in a straight line (figure A).
   d. Backing may be installed by notching the studs and fastening to its edges (figure B).
8. Layout and Spacing of Ceiling Joists
   a. Lay out spacing on double top plate (figure A).
   b. For the first wall, mark an R for the rafter on one side of the layout line and an X or a C on the other side.
   c. Layout letters on the opposite wall are determined by the type of ceiling joist construction.
      - When ceiling joists are lapped, place the R of the second wall on the opposite side of the line as from the first wall.
      - When ceiling joists are in-line, place the R on the same side of the line as the first wall and mark the X and C.
   d. The layout letters on the load-bearing decision must reflect the decisions made on the outside walls.

VIDEO #6
Section 7

9. Installing Metal Framing
   a. Snap chalk lines on floor.

b. Plumb up from the partition ends and snap lines on ceiling.
c. Fasten track to floor and ceiling (figure A).
d. At corners, extend one track to the end, then butt or overlap the other track.
e. Lay out the stud spacing and wall openings on the bottom track.
f. Install the first stud plumb and lay out top plate from this stud (figure B).

10. Framing a Door Opening for a Three-Piece Frame

The three-piece door unit may be installed later after the wall is framed.

a. Fasten full length stud on each side of the opening.
b. Cut a piece of track for use as a header (figure A).
c. Install the door frame as per manufacturer's instructions.

11. Framing a Door Opening for a One-Piece Frame

a. Place the studs on each side of the opening but do not fasten to the track yet (figure A).
b. Set the one-piece door frame in place using the studs interlocked with the tracks to help hold the door unit (figure B).
c. Fasten the bottom ends of the door jambs to the floor in the proper locations.
d. Fasten the studs to the door jambs and fasten the studs to the bottom track.
e. Plumb the door frame by movement in the top track and fasten to the top track with screws (figure C).
f. Install header and jack studs in the same manner as described previously (figure D).

Chapter 9 Scaffolds, Ladders, and Sawhorses

Instructor's Note

In this column, you will find a variety of lecture tips to aid student comprehension, notes on when to use PowerPoint slides to teach concepts, safety cautions for your students, and notes on when your class should view a video segment.

Appendix A lists corresponding numbered PowerPoint Presentation titles (see page 197)

Appendix B lists corresponding numbered Video titles and sections as applicable (see page 201)

Introduction

1. Scaffolds, sometimes called staging, raise the work area to a desired level, thereby allowing work to continue up to great heights.
2. Ladders are the means by which these raised work areas are accessed.
3. Sawhorses can create a raised surface from which to work.
4. These construction aids are involved with most job site accidents.
5. Contractors are carefully watched by safety organizations to ensure their job sites are safe places to work.
6. Workers must be responsible for their own safety, as well as that of those around them.
7. Safety is a group effort and all must comply in order for it to exist on the job.

OBJECTIVES

- Identify and describe the safety concerns for scaffolds.
- Erect and dismantle metal scaffolding in accordance with recommended, safe procedures.
- Follow a recommended procedure to inspect a scaffold for safety.
- Describe the recommended capacities of various parts of a scaffold.
- Construct a scaffold work platform.
- Identify and describe the components of a fall protection system.
- Describe the safety concerns for mobile metal tubular scaffolds.
- Build a safe staging area using roof brackets.
- Safely set up, use, and dismantle pump jack scaffolding.
- Describe the safe use of ladders, ladder jacks, stepladders, and sawhorses.

Glossary of Scaffolds, Ladders, and Sawhorses Terms

Cleat – a small strip of wood applied to support a shelf or similar piece

Competent person – designated person on a job site who is capable of identifying hazardous or dangerous situations and who has the authority to take prompt corrective measures to eliminate them

Crib – heavy wood blocks and framing used as a foundation for scaffolding

Erectors – workers whose responsibility it is to safely assemble scaffolding

Users – people who work on scaffolding

A. Scaffolds

1. Scaffolds are an essential element of construction, as they allow work to be performed at various elevations.
2. All workers on scaffolds must wear hard hats.
3. Occupational Safety and Health Administration (OSHA) (figure 9-1)
   a. OSHA reports that in construction, falls are the number one killer.
   b. Forty percent of those injured in falls had been on the job for less than one year.
   c. OSHA regulations on the fabrication of frame scaffolds are found in the code of Federal Regulations 1926.450,451,452.
4. Proof of training must be in writing.
   a. Employers are responsible for ensuring that workers are trained to erect and use scaffolding.
   b. One level of training is required for workers, such as painters, to work from the scaffold. A higher level of training is required for workers erecting, dismantling, moving, operating, repairing, maintaining, or inspecting scaffolds.
5. Competent Person
   a. Must be able to identify existing and potential hazards.
   b. Has the authority to take prompt corrective measures or eliminate such hazards.
   c. Has the authority to take corrective measures and stop work if need be to ensure scaffolding is safe to use.

Have students inspect, name parts of, assemble, disassemble, and store a metal tubular frame scaffold. Repeat on various surfaces, e.g, flat floor and sloped outdoor terrain with mud sills.

6. Metal Tubular Frame Scaffold
   a. Consists of manufactured end frames with folding cross braces, adjustable screw legs, base plates, platforms, and guardrail hardware (figure 9-2).
7. Metal Scaffold Components
   a. End frames (figure 9-3).
   b. Locking Devices (figure 9-4).
   c. Cross braces (figure 9-5).
   d. Base plates and mud sills (figure 9-6).
   e. Screw jack legs (figure 9-7).
   f. Plank (figure 9-8).
   g. Guardrail (figure 9-9).
   h. Top rail.
   i. Eye toeboard.
   j. Coupling pins (figure 9-10).
   k. Uplift protection (figure 9-11).
   l. Scaffold ladders (figure 9-12).
   m. Access ladders (figure 9-13).
   n. Side brackets and hoist arms (figures 9-14 and 9-15).
8. Scaffold Inspection
   a. Almost half of scaffold accidents involve defective scaffolds or defective scaffold parts.
   b. Visual inspection of scaffold parts should take place at least five times.
      - Before erection.
      - During erection.
      - During scaffold use.
      - During dismantling.
      - Before scaffold parts are put in storage.

c. Damaged parts should be red-tagged and removed from service.
d. Things to Look for During Inspection:
   - Broken/rusted welds.
   - Split, bent, or crushed tubes.
   - Cracks in the tube circumference.
   - Distorted members.
   - Damaged brace locks.
   - Lack of straightness.
   - Worn rivets or bolts on braces.
   - Split ends on cross braces.
   - Bent or broken clamp parts.
   - Damaged threads on screw jacks.
   - Damaged caster brakes.
   - Damaged swivels on casters.
   - Metal fatigue caused by temperature extremes.
   - Corrosion.
   - Leg ends filled with dirt or concrete.

9. Scaffold Capacity
   a. Safety margin – erectors and users of the scaffold must know the maximum intended load and the load-carrying capacities of the scaffold they are using.
   b. Scaffold coverings – if the scaffold is covered with weatherproofing, plastic, or tarps, the number of tie-ins attached to prevent overturning must be increased.
   c. Tie-ins – supported scaffolds with a ratio larger than 4-to-1 of the height-to-narrow-base width must be restrained from tipping by guying, tying, bracing, or equivalent means.
   d. Component loads – it is possible to overload the bottom legs of the scaffold without overloading the bearer, or top horizontal member, of any frame, and vice versa.
   e. Base plate – supported scaffold poles, legs, posts, frames, and uprights should bear on base plates and mud sills.
   f. Mud sills – made of 2 × 10 full thickness (figure 9-16).
   g. Unequal loading – the loads exerted on the legs of a scaffold are not equal (figure 9-17).
10. Scaffold Platforms
   a. Work area must be fully planked.
   b. No more than 1-inch gap between planks unless to fit around uprights. Any remaining gap must not exceed 9 1/2 inches.
   c. Platforms must be 18 inches wide, minimum.
   d. Maximum distance away from work: 14 inches.
   e. Overlap should not be less than 12 1/2 inches unless platforms are nailed together.
   f. Uncleated planks should extend over support, minimum of 6 inches and maximum of 12 inches.
   g. When fully loaded, wood planks must never deflect more than 1/60th of their span.
11. Scaffold Access (figure 9-19)
   a. A means of access must be provided to any scaffold platform that is 2 feet above or below a point of access (i.e., a hook-on or attachable ladder, a ramp, or a stair tower) and is determined by the competent person on the job.

   b. Scaffold users should never climb any end frame unless the manufacturer of that frame designated it to be used for access.

12. Scaffold Use
   a. Scaffolds must not be loaded in excess of their maximum intended load or rated capacities.
   b. Employees must not work on scaffolds covered with snow or ice, except to remove snow or ice.
   c. Makeshift scaffold devices such as boxes or barrels must not be used on the scaffold.
   d. Stepladders may be used on the scaffold platform if they are secured according to OSHA regulations.
13. Fall Protection
   a. Required when workers are working at heights above 10 feet.
   b. Guardrail System:
      - Top rail: 38 to 45 inches.
      - Mid rail.
      - Toe board.
   c. Personal Fall Protection System: (figure 9-20)
      - Harness.
      - Lanyard.
      - Rope grab.
      - Lifeline.
      - Anchor.
   d. OSHA recognizes that fall protection may not be possible for erectors.
   e. OSHA puts the responsibility of when to use fall protection, both for the user of the scaffold and the erector, on the competent person.
14. Falling Object Protection
   a. Workers must wear hard hats during the process of erecting a scaffold.
   b. OSHA lists toe boards as part of the falling object protection for workers below the scaffold.
15. Dismantling Scaffolds
   a. Check every scaffold before dismantling.
   b. If a hoist is to be used to lower the material, the scaffold must be tied to the structure at the level of the hoist arm.
   c. The erector should be tied off for fall protection.
   d. Start at the top and work in reverse order.
   e. Do not throw planks or material from scaffold.
   f. Building tie-ins and bracing can only be removed when the dismantling the process has reached that level or location on the scaffold.
   g. Never climb or access the scaffold by using the cross braces.
   h. As scaffold parts come off the scaffold, they should be inspected for wear or damage.
   i. Dismantled parts and materials should be organized, stacked, and placed in bins or racks out of the weather.
16. Mobile Scaffolds
   a. The rolling tower, or mobile scaffold, is generally not more than 20 feet in height (figure 9-21).

Have students inspect, name parts of, assemble, disassemble, and store a mobile scaffold tower.

b. Components are the same as those for the stationary frame scaffold, with the addition of casters (figure 9-22) and horizontal diagonal bracing.

c. The height of a rolling tower must never exceed 4 times the minimum base dimension.

d. When outriggers are used, they must be on both sides.

e. Casters must be locked when scaffold is stationary.

f. Casters typically have a load capacity of 600 pounds each, and the legs of a frame scaffold can hold 2000 to 3000 pounds each. Care must be taken not to overload the casters.

g. Never put cantilevered work platforms or side brackets on a mobile tower.

h. Mobile towers must have horizontal, diagonal, or gooser braces at the base to prevent racking of the tower during movement (figure 9-23).

i. Employees are not allowed to ride on the rolling tower scaffolds during movement unless the height-to-base width ratio is two-to-one or less.

j. Caster and wheel stems shall be pinned or otherwise secure.

B. Pump Jack Scaffolds

Erect pump jack scaffold according to OSHA regulations.

1. Consist of 4 × 4 poles, a pump jack mechanism, and metal braces for each pole (figures 9-24 and 9-25).
2. No more than two persons are permitted at one time between two supports.
3. Wood poles must not exceed 30 feet in height.
4. Braces must be installed at a maximum vertical spacing of not more than 10 feet.
5. In order to pump the scaffold past a braced location, temporary braces are used.

C. Roofing Brackets

Install roofing brackets.

1. Are used when the pitch of the roof is too steep for carpenters to work on without slipping (figure 9-26).
2. A metal plate at the top of the bracket has three slots in which to drive nails to fasten the bracket to the roof.
3. Applying roof brackets
   a. Apply roof brackets in rows at about 6 to 8 foot horizontal intervals.
   b. Nail the bracket over a joint or cutout in the tab of the shingle course below.
   c. Overlap staging planks same as wall scaffolds.
   d. A toe board of 1 × 6 or 1 × 8 is usually placed flat on the roof to protect new roofing from workers' boots (figure 9-27).

D. Ladders

Before raising an extension ladder, always check for overhead obstructions, particularly power lines.

1. Extension ladders
   a. Made of wood, aluminum, or fiberglass.
   b. Raising an extension ladder:
      - Place ladder's feet against a solid object.
      - Pick up other end.
      - Walk forward under ladder, pushing upward on each rung until the ladder is upright (figure 9-28).
      - Make sure the spring-loaded hooks are over the rungs on each side.
      - Move the base out until the distance from the wall is about 1/4 the vertical height (figure 9-29).
      - If the ladder is used to reach a roof or working platform, it must extend above the top support by at least 3 feet.

Erect and use extension ladders and stepladders, including ladder jacks.

VIDEO #4
Section 3

- Always face the ladder when climbing.
- Always grasp the rungs with both hands, making 3-point contact at all times (figure 9-30).

2. Stepladders
   a. Open legs fully so the brackets are straight and locked.
   b. Never work above the second step from the top.
   c. Do not use a folded stepladder that leans against a wall or object.
   d. Do not overreach (keep belt buckle between side rails).
   e. Remove tools and materials from ladder before moving it.
3. Ladder Jacks
   a. Ladder jacks are metal jacks installed on ladders to hold scaffold plank.
   b. No more than two persons should occupy any 8 feet of ladder jack scaffold at one time.
   c. Platform width must not be less than 18 inches.
   d. Planks must overlap the bearing surface by at least 10 inches (figure 9-31).

Construction Aids

1. Sawhorses (figure 9-32).
2. Job-made ladders (figure 9-33).

E. Scaffold Erection Procedure

1. Pre-assembly and Inspection
   - All workers must wear hard hats.
   - Inspect all components.
   - The foundation of a scaffold must be stable and sound.
   - Start erecting the scaffold at the highest elevation.
   - Scaffold must always be level and plumb.
2. Scaffold Assembly Procedures
   - Lay out location of base plates and screw jacks on mud sills (figure A).
   - Stand one of the end frames up and attach cross braces to each side.
   - Connect the other end of the braces to second end frame (figure B).
   - Plumb and level each frame (figure C).
   - Keep scaffold bases square with each other.
   - Place planks on top of end frames.
     1) Planking must meet OSHA requirements.
     2) Must extend over their end supports by at least 6 inches and not more than 12 inches (figure D).
     3) The cleat should be nailed to both ends of wood planks to prevent movement.
     4) Platform laps must be at least 12 inches.
     5) Minimum of two planks during erection.
     6) Full decking required when considered a working platform.
   - Access may be on end frame, if it is so designed, or an attached ladder (figure E).
   - Second-level frames may be hung over ends of first frames onto the coupling pins of the first level frames (figure F).
   - When lifting and placing frames, OSHA requires erector fall protection (body harness) when it is feasible and not a greater hazard to do so.

3. Post-Assembly Checklist
   - Fully decked platform, guardrail system, personal fall-arrest system, before it can be turned over to scaffold users.
   - If scaffold is higher than four times its minimum base dimension, it must be restrained (figure G).
   - Plumb, level, and square.
   - All legs are on base plates or screw jacks and mud sills (if required).
   - Toeboards and/or screening installed as needed.
   - End and/or side brackets are fully secured.
   - Access units are correctly installed.
   - Must be inspected by a competent person at the beginning of each work shift and after any occurrence, such as high wind or a rainstorm.

Chapter 10 Roof Framing

Instructor's Note

In this column, you will find a variety of lecture tips to aid student comprehension, notes on when to use PowerPoint slides to teach concepts, safety cautions for your students, and notes on when your class should view a video segment.

Appendix A lists corresponding numbered PowerPoint Presentation titles (see page 197)

Appendix B lists corresponding numbered Video titles and sections as applicable (see page 201)

Introduction

1. Due to the complexity of the task, roof layout and member cutting is often performed by the most experienced carpenters on the job.
2. Learning roof framing takes patience, practice, and a willingness to persevere.
3. Being a good roofing carpenter begins with knowing the names of all the components.
4. Knowledge of diverse roof types and the terms used in their construction is essential.
5. Knowledge of straightforward math and geometry in roof framing is very helpful in making accurate measurements and cuts.

OBJECTIVES

- Describe several roof types.
- Define the various roof framing terms.
- Identify the members of gable, gambrel, hip, intersecting, and shed roofs.
- Lay out a common rafter and erect a gable roof.
- Lay out and install gable end studs.
- Lay out a hip rafter and hip jack rafters.
- Lay out a valley rafter and valley jack rafters.
- Describe and perform the safe and proper procedure to erect a trussed roof.
- Apply roof sheathing.
- Estimate the quantities of materials used in a roof frame.

Glossary of Roof Framing Terms

Cheek cut – a compound miter cut on the end of certain roof rafters

Fascia – a vertical member of the cornice finish installed on the tail end of rafters

Gable roof – a type of roof that pitches in two directions

Gambrel roof – a type of roof that has two slopes of different pitches on each side of center

Hip jack – a rafter running between a hip rafter and the wall plate

Hip rafter – extends diagonally from the corner of the plate to the ridge at the intersection of two surfaces of a hip roof

Hip-valley cripple jack rafter – a short rafter running parallel to common rafters, cut between hip and valley rafters

Intersecting roof – the roof of irregularly shaped buildings; valleys are formed at the intersection of the roofs

Lateral – a direction to the side at about 90-degrees

Lookout – horizontal framing pieces in a cornice, installed to provide fastening for the soffit

Mansard roof – a type of roof that has two different pitches on all sides of the building, with the lower slopes steeper than the upper

Rake – the sloping portion of the gable ends of a building

Shed roof – a type of roof that slopes in one direction only

Tail cut – a cut on the extreme lower end of a rafter

Valley – the intersection of two roof slopes at interior corners

Valley cripple jack rafter – a rafter running between two valley rafters

Valley jack rafter – a rafter running between a valley rafter and the ridge

Valley rafter – the rafter placed at the intersection of two roof slopes in interior corners

SLIDE 19

A. Roof Framing Terms

1. Rafter
   a. The sloping structural members of a roof frame.
   b. Most rafters are common rafters.
   c. Other types of rafters include hip, valley, and their respective jack rafters.
2. Span
   a. The span of a roof is the horizontal distance covered by the roof.
   b. The span is usually the width of the building being measured.
3. Run – typically 1/2 the span.
4. Rise – the total rise is the total vertical distance that the roof rises.
5. Ridge – secures the upper end of rafters.
6. Seat Cut or Bird's Mouth
   a. The notch cut near the lower end of a rafter.
   b. Where the rafter will be fastened to the wall.
7. Line Length – the hypotenuse (longest side) of a right triangle formed by the rafter.
8. Unit Triangle
   a. The small right triangle found on the set of prints for the building.
   b. Its run and its rise.
9. Unit Run
   a. The unit run is 12 inches (one foot).
   b. The unit run of hip and valley rafters is 16.97 inches because the typical run of hip and valley rafters project into the building at a 45-degree angle with the plates.
10. Unit Rise – the distance that the roof rises vertically for every unit of run.
11. Unit Length – the length of rafter necessary to cover one unit of run.
12. Pitch – the ratio of rise to span of a roof.
13. Plumb Line
    a. Any line on the rafter that is vertical when the rafter is in position.
    b. Marked along the edge of the square where the inch ruler is located (figure 10-3).
14. Level Line – any line on the rafter that is horizontal when the rafter is in position.

SLIDE 20

B. Roof Types

1. Several roof styles are in common use (figure 10-1).
2. The most common roof style is the gable roof, where two sloping roof surfaces meet at the top.
3. The shed roof slopes in one direction.
4. The hip roof slopes upward to the ridge from all walls of the building.
5. An intersecting roof is where two roofs intersect, forming valleys.
6. The gambrel roof is a variation of the gable roof.
7. The mansard roof is a combination of the gambrel and hip roofs.
8. The butterfly roof is an inverted gable roof.

C. Gable Roof

1. This is the simplest roof to frame (figure 10-4).
2. Gable roofs may have two different runs on each side of the ridge, such as the saltbox roof (figure 10-5).
3. Only common rafters need to be laid out (figure 10-10). Step-by-step instructions for laying out a common rafter are in the procedures section of this chapter.
   a. Plumb cut
   b. Seat cut
   c. Tail cut
4. There are two methods of determining rafter length.
   a. Rafter tables, which are stamped onto one side of most framing squares (figure 10-7). Framing square gauges are attached to the square to simplify repetitive alignments (figure 10-9).
   b. Step-off method – the rafter stock is stepped off for each unit of run until the desired number of units are stepped off (figure 10-11).
5. Ridgeboard Layout
   a. Transfer rafter layout from the plate to the ridge board.
   b. Joints in the ridgeboard should be centered on a rafter.
   c. If there is an overhang, add the necessary amount on both ends.
6. Erecting the Gable Roof
   a. Rafters are usually installed after the ceiling joists are installed (figure 10-14).
   b. Top of rafter is nailed to ridgeboard.
   c. Seat cut is nailed to wall plate and to the sides of the ceiling joists.
   d. Step-by-step instructions on erecting the gable roof frame are in the procedures section of this chapter.
7. Rafter Tails
   a. Some carpenters prefer to cut the rafter tails to length after they are installed (figure 10-15).
8. Rake Overhang
   a. Lookouts support the rake rafter (figure 10-16).
   b. Care should be taken to be sure the rake rafter is straight.
9. Roof Openings
   a. Small openings for skylights and chimneys may be created by using headers in a similar fashion as with ceiling joists (figure 10-17).

D. Gable Studs

1. The bottom ends are cut square and toenailed to the top of the nail plate.

2. Top ends fit snugly against the bottom edge and inside face of the end rafter (figure 10-18).
3. Studs should be positioned directly above end wall studs.
4. They have a common difference in length (figure 10-19).
   a. To find the common difference in length of the gable studs, multiply the stud spacing in feet by the unit rise of the roof.
5. Step-by-step instructions on installing gable studs are in the procedures section of this chapter.

SLIDE 21

E. Hip Roof

1. To frame a hip roof, it is necessary to lay out common rafters, a ridge, hip rafters, and hip jack rafters (figure 10-20).
2. Hip Rafter
   a. The hip rafter is run at a 45-degree angle from the plate.
   b. The unit run for a hip is the diagonal of the square formed by the unit run of a common rafter and the exterior walls.
   c. This figures to be 16.97 inches or 17 inches for all practical purposes (figure 10-21).
   d. Steps to lay out a hip rafter are similar to a common rafter.
   e. Shortening the hip rafter is 1/2 the 45-degree thickness of the ridgeboard (figure 10-22).
   f. Cheek cuts
      - The ridge cut of a hip rafter is a compound angle called a cheek cut or side cut.
      - A single cheek cut or a double cheek cut may be made on the hip rafter according to the way it is framed at the ridge (figure 10-23).
      - The tail cut may also be cut with cheek cuts to allow the tail to fit against the fascia.
   g. Backing or dropping the hip is done to accommodate alignment of the hip rafter with the plane of the roof (figure 10-24).
3. Hip Jack Rafter
   a. Hip jack rafter is a common rafter that is shortened with cuts similar to hip rafters.
   b. The seat cut and tail cut are the same as that of the common rafter (figure 10-25).
   c. Hip jack rafters are framed against the side of the hip rafter.
   d. Lengths may be found using the common difference found in the rafter tables stamped on the framing square (figure 10-26).
   e. Jack rafter has a single cheek cut where it meets the hip rafters.
4. Hip Roof Ridge Board Length

SLIDE 22

   a. Found by subtracting the width of the building from the length of the building.
   b. If the hip rafters are framed against common rafters, the ridge length is increased at each end by 1/2 the thickness of the common rafter stock.
   c. If the hip rafters are framed against the ridge, increase the length of each end by half the 45-degree thickness of the hip rafter plus half the ridge board thickness (figure 10-27).

F. Valley Rafters

1. Valley rafters form the intersection of the slopes of two roofs. If the heights of the roofs are different, two kinds of valley rafters are required.

2. Identification of Valley Roof Frame Members
    a. Supporting valley rafter runs from the plate to the ridge of the main roof.
    b. Shortened valley rafters run from the plate to the supporting valley rafter.
    c. Valley jack rafters run from the ridge to the valley rafter.
    d. Hip-valley cripple jack rafter runs between a hip rafter and a valley rafter.
    e. Valley cripple jack rafter runs between the supporting and shortened valley rafter (figure 10-28).
3. Rules to remember concerning the layout of so many different kinds of rafters:
    a. Hip and valley rafters are similar.
    b. Common rafters and all jack rafters are similar.
    c. The length of any kind of rafter can be found from its run and its unit length.
    d. The amount of shortening is always measured at right angles to the plumb cut.
    e. The method for laying out cheek cuts for all rafters is similar.
4. Supporting Valley Rafter Layout
    a. Similar to hip rafters.
    b. Unit of run is 17 inches.
    c. The total run of the supporting valley is the run of the common rafter of the main roof, called the major span.
    d. Its unit length is found on the rafter tables.
    e. A single cheek cut is made at the ridge.
    f. The rafter is shortened by half the 45-degree thickness of the ridge board.
    g. Step-by-step instructions on valley rafter layout are in the procedures section of this chapter.
5. Shortened Valley Layout
    a. The length of the shortened valley is found by using the run of the common rafter for the smaller roof, called the minor span.
    b. The shortening is 1/2 the thickness of the supporting valley rafter (figure 10-29).
6. Valley Jack Layout

SLIDE 23

    a. Length can be found by multiplying its run by its unit length.
        - The total run of a valley jack rafter is equal to the run of the common rafter minus the horizontal distance from the inside corner of the building (figure 10-30).
    b. Ridge cut is the same as a common rafter.
    c. A single cheek cut is shortened by 1/2 the 45-degree angle thickness of the valley rafter stock.
    d. Length of all other valley jack rafters is found using the common difference found on the rafter tables.
7. Hip-valley Cripple Jack Rafter Layout
    a. The run is equal to the distance between the seat cuts of the hip and valley rafters, measured along the wall plate (figure 10-31).
    b. Determine the length using the unit length for the common rafter.
    c. All hip-valley cripple jacks cut between the same hip and valley rafters are the same length.
8. On each end, shorten by 1/2 the 45-degree thickness of the hip and valley rafters (figure 10-32).

G. Shed Roof

1. A shed roof rafter is similar to a common rafter.
2. It may require two seat cuts instead of one (figures 10-33 and 10-34).

H. Dormers

1. Dormers have either gable or shed roofs.
2. Rafters on both sides of the framed openings in the roof are doubled.
3. Some dormers have their front walls directly over the exterior wall below (figure 10-35).

I. Other Roof Framing

1. Fitting a valley jack rafter to roof sheathing
   a. Some intersecting roofs are built after the main roof has been framed and sheathed.
   b. Steps to lay out the seat cut of the valley jack rafter to fit against roofs of the same and different inclines are shown in figure 10-36.
   c. Lay out of the ridge cut against roof sheathing (figure 10-37).
2. The top ends of shed roof rafters are occasionally fitted against a roof of a different and steeper incline.

VIDEO #4
Section 1

Never cut any webs or chords of a truss unless directed by an engineer. Also, installing trusses can be very dangerous. Lives have been lost installing trusses improperly.

J. Roof Trusses

1. General
   a. Because of their design, trusses can support a roof over wide spans (figure 10-39).
   b. Trusses eliminate the need for load bearing partitions below.
   c. The roof is also framed in much less time.
2. Truss Design
   a. Upper chords act as rafters.
   b. Lower chords serve as ceiling joists.
   c. Joints are fastened securely with metal or wood gusset plates (figure 10-40).
   d. Most common design for residential construction is the fink truss (figure 10-41).
   e. Other truss shapes are designed to meet special requirements (figure 10-42).

VIDEO #4
Section 4

The erection and bracing of a trussed roof is a critical stage of construction. Failure to observe recommendations for erection and bracing could cause a collapse of the structure. This could result in loss of life, serious injury, or loss of time and material.

3. Overview of Truss Roof Erection
   a. Should be loaded and stored on a flat, dry surface.
   b. A drawing of each truss is provided that outlines important installation points (figure 10-43).
   c. Trusses are temporarily and permanently braced to tie them together.
      - Permanent bracing remains in place after construction is completed and is vital to the strength of the roof structure.
      - Temporary bracing is designed to create a safe work environment during truss erection.
   f. Trusses are installed one at a time by lifting, fastening, and bracing in place.
   g. Installing the first truss
      - The end truss is usually installed first and braced securely in a plumb position.
      - May be achieved by bracing to securely anchored stakes driven into the ground or bracing to inside floor under the truss (figure 10-44).
      - Bracers should be located directly in line with continuous top chord lateral bracing, which will be installed later.

VIDEO #4
Sections 7, 8

VIDEO #4
Section 8

h. Temporary Bracing of Trusses
   - As each truss is set in place, it is nailed to the plate and bracing (figure 10-45).
   - Temporary bracing applied according to manufacturer's recommendation.
   - Temporary bracing should be no less than 2 × 4 lumber.
   - 2 × 4s should be fastened with two 16d duplex or common nails at every intersection.
   - Spacing of trusses should be accurately maintained.
   - Continuous lateral bracing should be installed within 6 inches of the ridge and in rows of 8 to 10 feet intervals between the ridge and the wall plate (figure 10-46).
   - A diagonal brace should be set at a 45-degree angle between the rows of lateral bracing (figure 10-47).
   - Bracing the web member plane, diagonals are placed at right angles to the trusses from the top to the bottom chords (figure 10-48).

i. Bracing the Plane of the Bottom Chord
   - Continuous lateral bracing for the full length of the building must be applied.
   - Bracing should be nailed to the top of the bottom chord at intervals of 8 to 10 feet along the width of the building.
   - Diagonal bracing should be installed at each end of the building (figure 10-49).

j. Permanent Bracing
   - Designed by the engineer.
   - Top chord permanent bracing is often provided by the roof sheathing.
   - Web bracing may be lateral bracing or web stiffeners.

k. Framing Openings
   - Openings in the roof or ceiling for skylights or access ways must be framed within or between the trusses.
   - Chords, braces, and webs of a truss system should never be cut or removed unless directed by an engineer.

VIDEO #4
Section 10

K. Roof Sheathing

1. Rated panel sheathing is laid with face grain running across the rafters.
2. End joints are made on top of rafters/top chord and staggered.
3. Nails are spaced 6 inches apart on the ends and 12 inches apart on intermediate supports (figure 10-51).
4. Blocking, tongue-and-groove edges, or panel clips must be used in between supports (figure 10-52).

L. Estimating

1. Common Rafters for Gable Roof
   a. Divide the length of the building by the spacing of the rafters.
   b. Add 1 as a starter.
   c. Multiply the total by 2.
2. Common Rafters with Hip or Valley Rafters
   a. Use the same procedure as for gable roof, then add 2 rafters for each hip and valley rafter.
3. Hip and Valley
   a. Count the number of hips and valleys.

4. Ridgeboard for Gable Roof
    a. The length of the building plus the length of overhangs.
5. Ridgeboard for Hip Roof
    a. Subtract the width of the building from the length of the building.
    b. Add 4 inches.
6. Gable End Studs
    a. Width of the building divided by the on-center spacing plus 2 extra, enough for 2 gable ends.
7. Gable Roof Fascia
    a. Double the number of gable ridgeboards.
8. Hip Roof Fascia
    a. Twice the building length plus twice the building width.
    b. Add 8 times the rafter projection.
9. Trusses
    a. Building length divided by on-center spacing.
    b. Add 1 to start.
10. Bracing Material
    a. Building width divided by 4 gives the number of rows of top, bottom, and web bracing.
    b. Add extra for ground bracing the first truss.
11. Sheathing Gable Roof
    a. Length of rafter times length of ridge.
    b. Divide this number by 32.
    c. Round up to nearest whole number.
    d. Equals sheathing for one side.
12. Sheathing a Hip Roof
    a. Calculate as though a gable roof.
    b. Add 5%.
13. Gable End Sheathing
    a. Multiply the total rise by the rafter run.
    b. Multiply this number by 2.
    c. Divide by 32.
    d. This will allow material for 2 gable ends.

SLIDE 24

M. Procedures

1. Common rafter layout
    a. Crown the board.
    b. Adjust square until outside edges are at specified unit rise and unit run.
    c. Mark rafter along outside edge of tongue (figure A).
    d. Shorten the rafter half the thickness of the ridgeboard at plumb cut (figure B).
    e. Determine rafter line length (figure C).
    f. Measure from first ridge plumb cut line along the edge of the rafter.
    g. Make a plumb line for the seat cut (figure D).
    h. Draw seat cut (figures E and F).
    i. Lay out rafter overhang (figure G).
    j. A level cut may be made to remove any portion of rafter extending below the fascia (figure H).

SLIDE 25

Check the straightness of the top plates periodically as rafters are installed. If wall bracing is not adequately installed or nailed, the walls will bow out as the rafters are installed, causing the ridge to sag.

2. Erecting the Gable Roof Frame
    a. Lay ridgeboard on top of the work platform.
    b. Select four straight pieces for rake rafters.
    c. Fasten rafter to each end of the section of ridge board.
    d. Raise the ridgeboard and the two rafters into position and fasten at seat cut (figure A).
    e. Install two opposing rafters on other side.
    f. Plumb and brace this section (figure B).
    g. Install remaining rafters (figure C).
3. Installing Gable Studs
    a. Lay out the top plate.
    b. Plumb a stud from the layout mark on the plate. Mark along top and bottom edge of rafter (figure A).
    c. Determine the common difference in length of studs.
    d. Lay out and cut studs (figure B).
    e. Install studs.
    f. End ceiling joists are nailed to the inside edges of the studs.

SLIDE 26

4. Hip Rafter Layout
    a. Select stock.
    b. Crown stock.
    c. Mark plumb line at one end.
    d. Hold the tongue of the square at the unit rise and the blade of the square at 17 inches.
    e. Shorten rafter at plumb cut 1/2 the 45-degree thickness of the ridge (figure A).
    f. From the top of this second plumb line, square a line over the top edge of the rafter. Mark the midpoint of that line.
    g. Lay out cheek cuts (figure B). Draw a third plumb line through this measurement.
    h. Draw line over the top of the rafter from the third plumb through the midpoint of the squared line of the second plumb line. This new line and the third plumb line become the cut lines of the ridge cut.
    i. Repeat for second cheek cut if necessary.
    j. Determine length of the hip rafter by using the rafter tables.
    k. Measure and mark length of the rafter (figure C).
    l. Hip seat cut has the same height above the seat cut as for the common rafter (figure D).
    m. Determine the amount the seat cut level line is raised (figure E).
    n. Calculate hip tail length (figure F).
    o. Lay out and cut the tail (figure G).

SLIDE 27

5. Jack Rafter Layout
    a. The first jack rafter length is calculated by multiplying run by unit length (figure A).
    b. Place the common rafter pattern on a piece of stock and mark the tail, seat cut, and seat cut plumb line.
    c. Mark length of rafter by measuring from the seat cut plumb line.
    d. Shorten the rafter for the thickness of the hip using the same procedure as for hip shortening at the ridge (figure B).

e. Lay out cheek cuts (figure C).

f. Remaining jack lengths may be determined by adding or subtracting a common distance in length found in the rafter tables.

SLIDE 28

6. Valley Rafter Layout

a. Select stock.

b. Crown stock.

c. Mark a plumb line at one end.

d. Shorten rafter by measuring at right angles from the plumb line 1/2 the 45-degree thickness of the ridge (figure A).

e. Lay out cheek cuts (figure B).

f. Determine the length of the valley rafter by using the same procedure as for the hip rafter.

g. Measure and mark the length of the rafter (figures C and D).

h. Lay out a deeper seat cut plumb line to allow for the thickness of the valley rafter (figure E).

i. The valley tail is usually not needed.

VIDEO #4
Sections 4, 7, 8

Trusses can be very dangerous to install. Workers must be alert at all times. Watch for proper nailing and order of assembly. Make sure bracing is securely and appropriately nailed.

7. Truss Installation

a. Check the walls for plumb and stiffness of temporary wall bracing.

b. Read and understand the prints and installation guide provided for proper location of trusses and bracing.

c. Secure brace bases to floor or stakes driven in ground (figure A).

d. Raise the first truss into place and raise securely.

e. Raise the next four trusses into place individually. Secure all five trusses together with lateral, web, and diagonal bracing.

f. Install remaining trusses and bracing. Install additional diagonal and permanent bracing as required.

Chapter 11 Windows and Doors

Instructor's Note

In this column, you will find a variety of lecture tips to aid student comprehension, notes on when to use PowerPoint slides to teach concepts, safety cautions for your students, and notes on when your class should view a video segment.

Appendix A lists corresponding numbered PowerPoint Presentation titles (see page 197)

Appendix B lists corresponding numbered Video titles and sections as applicable (see page 201)

Introduction

1. Logically, the installation order of house wrap, windows, and doors, and then siding must be followed.
2. Interior and exterior doors, like windows, are manufactured in millwork plants in a wide range of styles and sizes.
3. Most doors come prehung in frames, complete with exterior casings applied, and ready for installation. Doors come in many styles and sizes.
4. Care must be taken to maintain the easy operation and weathertightness of the window and door units.
5. Safety concerns remain a top priority as many windows are installed from scaffolds and often stepladders are used. Create safe work habits.

OBJECTIVES

- Describe the most popular styles of windows and name their parts.
- Select and specify desired sizes and styles of windows from manufacturers' catalogs.
- Install various types of windows in an approved manner.
- Name the parts of and set a prehung door frame.
- Describe the standard designs and sizes of doors and name their parts.
- Fit and hang a door to a preexisting opening.
- Install locksets in doors.
- Install bypass, bifold, and pocket doors.

Glossary of Window and Door Terms

Astragal – semicircular molding often used to cover a joint between doors

Back miter – an angle cut starting from the end and coming back on the face of the stock

Bay window – a window, usually three-sided, which projects out from the wall line

Casing – Molding used to trim around doors, windows, and other openings

Deadbolt – door-locking bolt operated by a key from the outside and by a handle or key from the inside

Double-acting – doors that swing in both directions; also the hinges used on these doors

Escutcheon – protective plate covering the knob or key hole in doors

Extension jambs – strips of wood added to window jambs to bring the jamb edge flush with the wall surface in preparation for casing

Glazing – the act of installing glass in a frame

Hopper window – a type of window in which the sash is hinged at the bottom and swings inward

Housewrap – type of building paper with which the entire exterior sidewalls of a building are covered

Insulated glass – multiple panes of glass fused together with an air space between them

Light – a pane of glass or an opening for a pane of glass muntins – vertical or horizontal members that divide a piece of glass into several smaller pieces; also, the vertical member that divides the panels of a door

Low-emissivity glass (Low E) – a coating on double-glazed windows designed to raise the insulating value by reflecting heat

Molding – decorative strips of wood used for finishing purposes

Mullion – a vertical dividing member between window units

Rail – the horizontal member of a frame

Sash – that part of a window into which the glass is set

Stile – the outside vertical members of a frame, such as in a paneled door

Strike plate – thin metal plate installed where the latch bolt of a door touches the jamb

Weatherstripping – narrow strips of thin metal or other material applied to windows and doors to prevent the infiltration of air and moisture

Wind – a defect in lumber caused by a twist in the stock from one end to the other; also, a twist in anything that should be flat

VIDEO #5
Section 1

A. Windows

1. Windows are one of many types of millwork (figure 11-1). Windows are made with wood, aluminum, steel, and vinyl. Windows made with the wood parts encased in vinyl are called vinyl-clad windows.
2. Parts of a Window
   a. Sash Parts
      - Vertical edge members of the sash are called stiles.
      - Top and bottom horizontal members are called rails.
      - The pieces of glass in a sash are called lights.
      - Small strips of wood that divide the glass into smaller lights are called muntins (figure 11-2).
      - Many windows come with false muntins called grilles (figure 11-3).

   b. Window Glass
      - The installation of glass in a window sash is called glazing.
      - Single strength (SS) glass is about 3/32 inch thick. It is used for small lights of glass.
      - For larger lights, double strength (DS) glass about 1/8 inch thick may be used.
      - Safety Glass – when broken at any point, the entire piece immediately disintegrates into a multitude of small granular pieces.
      - Insulated glass consists of two or sometimes three layers of glass separated by a sealed air space 3/16 to 1 inch in thickness (figure 11-4).
      - Solar Control Glass – the R-value of windows may also be increased by using special solar control glass. An invisible, thin, metallic coating is bonded to the air space side of the inner glass. This lets light through, but reflects heat.

c. The Window Frame
   - Usually comes with the exterior trim applied.
   - Consists of several distinct parts (figure 11-6).

d. The Sill
   - The bottom horizontal member of the window frame, usually at an angle to shed water.
   - Its bottom side usually is grooved so a weathertight joint can be made with the wall siding.

e. Jambs
   - The vertical sides of the window frame are called side jambs.
   - The top horizontal member is called a head jamb.

f. Extension jambs are applied to the inside edge of the jambs of the window frame (figure 11-7).

VIDEO #5 Section 2

g. Casings
   - Window units usually come with exterior casings applied.
   - The interior casing is applied after installation.
   - The casing covering the joint is called a mullion casing.

h. Window Flashing
   - Sometimes called a drip cap.
   - Bent to fit over the head casing and up against the exterior wall (figure 11-8).
   - The flashing sheds water over the top of the casing.

VIDEO #5 Section 1

i. Protective Coatings
   - Wood window units are primed.
   - Vinyl-clad.

j. Screens
   - Optional accessories.
   - Mesh is usually plastic or aluminum.

VIDEO #5 Section 1

3. Types of Windows

a. Fixed Windows
   - Fixed windows consist of a frame in which a sash is fitted in a fixed position.
   - They are manufactured in many shapes.
   - May be assembled or combined with other types of windows.

b. Single- and Double-Hung Windows
   - The single-hung window consists of two sashes where the upper one is fixed and the bottom one slides vertically.
   - The double-hung window also has two sashes, but both sashes slide vertically by each other in separate channels of the side jambs (figure 11-10).
   - Figure 11-11 shows double-hung windows used in a bay window unit.

c. Casement Windows
   - Consist of sash hinged at the side.
   - Swing outward by means of a crank or lever.

d. Sliding Windows
   - Have sashes that slide horizontally in separate tracks located on the header jamb and sill (figure 11-13).

e. An awning window unit consists of a frame in which a sash hinged at the top swings outward by means of a crank or lever.

f. The hopper window is hinged at the bottom and swings inward (figure 11-14).

g. Skylight and Roof Windows

- Skylights provide light only.
- Roof windows contain operating sashes to provide light and ventilation (figure 11-15).

CAUTION House wraps are slippery. They should not be used in any application where they will be walked on.

VIDEO #5 Section 3

4. Window Installation

a. Before windows are installed the exterior walls are covered with a building housewrap.

b. The numbers or letters found in the floor plan identify the window in the window schedule.

c. Use galvanized casing or roofing nails spaced about 16 inches apart.

d. Nail length depends on the thickness of the casing and sheathing.

e. Nails are driven through the flange instead of the casing (figure 11-19).

f. Clearance should be left for caulking around the entire perimeter between the window and the masonry (figure 11-20).

B. Doors

VIDEO #5 Section 6

Take students for a walk through school building and identify various door sizes, swings, backsets, and hardware.

1. Door Styles and Sizes

a. Practically all exterior entrance doors are manufactured in a thickness of 1 3/4 inches, in widths of 2′8″ and 3′0″, and in heights of 6′8″ and 7′0″.

b. Most interior doors are manufactured in 1 3/8 inch thickness. Most interior doors are manufactured in 6′8″ heights. Door widths range from 1′0″ to 3′0″ in increments of 2 inches.

c. A swinging door is designated as being right-hand or left-hand, depending on the direction it swings (figure 11-21).

d. Flush Doors

- An exterior flush door has a smooth, flat surface of wood veneer or metal. It has a framed, solid core (figure 11-22).
- Interior flush doors are made with solid or hollow cores (figure 11-23).

e. Panel Doors (figures 11-24 and 11-25)

- A panel door consists of a frame that surrounds panels of solid wood and glass, or louvers.
- The outside vertical members are called stiles.
- Horizontal frame members are called rails.

f. High-style doors are highly crafted designer doors (figure 11-27).

g. French doors may contain from one to fifteen lights of glass (figure 11-28).

h. Louver doors are made with spaced horizontal slats called louvers used in place of panels (figure 11-29).

i. Café doors are short panel or louver doors (figure 11-30).

j. Methods of Door Operation

- Doors are also identified by their method of operation.
- Doors either swing on hinges or slide in tracks.
- Swinging doors
    1. Are hinged on one edge (figure 11-31).
    2. May be single-acting doors.
    3. May be double-acting doors.
- Patio doors units normally consist of two or three full glass panels completely assembled in a frame. Typically, only one of the panels will move as a sliding or swinging glass door.

- Double doors are fitted and hung with an astragal (a molding designed to cover the joint between double doors) between them for weathertightness (figure 11-33).
- Bypass doors ride in a track mounted on the header jamb of the door frame and have rollers attached so that they slide by each other (figure 11-34).
- Pocket door is opened by sliding it sideways into the interior of the partition (figure 11-35).
- Bifold doors operate in a folding fashion on closet and similar type openings (figure 11-36). Bifold doors consist of pairs of doors hinged at their edges.

5. Parts of a Door Frame
   a. Terms given to members of an exterior door frame are the same as similar members of a window frame.
   b. Sills
      - Located at the bottom of the frame.
      - The shape of a wood door sill for an inward swinging door is different from that for an outward swinging door (figure 11-40).
      - Most sills are adjustable for exact fitting between the sill and door (figure 11-41).
   c. Jambs
      - Square-edge pieces of stock to which door stops are applied.
      - Several jamb widths are available for different wall thicknesses.
   d. Casings
      - The molding applied around the outside edges of the jamb.
      - Exterior casing is made with thicker material than interior (figure 11-42).
      - Interior casing is usually the same shape throughout the house.

VIDEO #5
Section 7

6. Setting a Prehung Door Frame
   a. A prehung single-acting, hinged door unit consists of a doorframe with the door hinged and casings installed to one side.
   b. Prehung units are available in several jamb widths to accommodate various wall thicknesses.
   c. Prehung units have split jambs that are adjustable for varying wall thicknesses (figure 11-44).
7. Setting Door Frames in Masonry Walls
   a. Head jamb is checked for level.
   b. The bottom ends of the side jambs are secured in place.
   c. The side jambs are then plumbed in a sideways direction and braced in position.
   d. The frame is plumbed and braced at a right angle to the wall (figure 11-45).
   e. Check the frame to see if it has a wind (figure 11-46).
8. Fitting a Door to a Door Frame
   a. Doors are made with a beveled edge that allows the door to open without touching the jamb (figure 11-47).
   b. The hinge leaves are recessed in flush with the door edge and only partway across.
   c. The recess for the hinge is called a hinge gain or hinge mortise.
9. Installing Exterior Door Locksets
   a. Cylindrical locksets are the most commonly used type (figure 11-49).

b. Lever handles are locksets likely to be used by handicapped persons or in other situations (figure 11-50).

c. Deadbolt locks are used for both primary and auxiliary locking of doors (figure 11-51).

d. Escutcheons are decorative plates of various shapes that are installed between the door and the lock handle or knob (figure 11-52).

e. A privacy lock is often used on bathroom and bedroom doors. It is locked by pushing or turning a button on the room side.

f. The passage lockset has knobs on both sides that are turned to unlatch the door. This lockset is used when locking the door is not necessary.

10. Garage Doors

a. The rigid one-piece unit swings out at the bottom. It then slides up and overhead.

b. The sectional type has hinged sections. These sections move upward on rollers and turn to a horizontal position overhead.

c. A rolling steel door consists of narrow metal slats that roll up on a drum installed above the opening.

C. Procedures

VIDEO #5
Sections 2, 3

Install a window, including housewrap.

1. Installing Windows

a. Install housewrap.
- Make sure the sheet is straight (figure A).
- Fasten every 12 to 18 inches.
- Overlap all joints by at least 3 inches.
- Secure them with housewrap tape.
- On horizontal joints, the upper layer should overlap the lower layer.
- Make cuts in the housewrap from corner to corner of rough openings (figure B).

b. Place window in the opening.

c. Close and lock the sash (figure C).

d. Center the unit in the opening on the rough sill.

e. Level the window sill (figure D).

f. Caulk the backside of the casing or nailing flange.

g. Plumb the side and nail the unit along the sides and top.

h. Check that the sash operates properly.

i. Flash the head casing (figure E).
- Cut the flashing to length with tin snips.
- Lap joint about 3 inches.
- Slip the flashing behind the wrap and on top of the head casing.
- Secure with fasteners.
- Refasten the house wrap.

SLIDE 29

VIDEO #7
Section 5

Have students install a prehung door.

2. Hanging Interior Doors

a. Setting a Prehung Door Frame.
- Remove the protective packing from the unit (figure A).
- Leave the small fiber shims between the door and jambs to help maintain this space.
- Center the unit in the opening, so the door will swing in the desired direction.
- Level the head jamb (figure B).

- Plumb the hinge side jamb of the door unit (figure C).
- Open the door and move to the other side (figure D). Shim as needed.
- Check the operation of the door.
- Finish nailing the casing and install casing on the other side of the door (figure E).

b. Fitting a Door to a Frame

Have students mortise a door for hinges and bore for lockset. A 3′ piece of 2″ × 8″ may be substituted for a door if doors are not available.

- Check the door for its beveled edge.
- Lightly mark the location of the hinges on the door. Nine inches down from the top and 13 inches up from the bottom, as measured to the center of the hinge. A middle hinge is centered between the two (figure A).
- Check the opening frame for level and plumb.
- Plane the door edges so the door fits onto the opening with an even joint of approximately 3/32 inch between the door and the frame on all sides (figure B).
- Place the door in the frame. Shim the door so the proper joint is obtained along all sides (figure C).
- Remove the door from the frame.
- Score the hinge lines, taking care not to split any part of the door (figure D).
- Press the hinge leaf into the gain joint. It should be flush with the door edge and install screws.
- Place the door in the opening and insert the hinge pins.
- Apply the door stops to the jambs (figure E).

c. Installing Bypass Doors

Have students install bypass and/or bifold doors. If entire installation is not possible, at least have them remove doors and reinstall doors into existing hardware.

- Cut the track to length. Install it on the header jamb (figure A).
- Install pairs of roller hangers on each door.
- Mark the location and bore holes for door pulls (figure B).
- Hang the doors by holding the bottom outward (figure C).
- Test the door operation and the fit against side jambs.
- Mark the location and fasten the floor guide (figure D).

d. Installing Bifold Doors

- Cut the track to length. Fasten it to the header jamb with screws provided in the kit.
- Locate the bottom pivot sockets. Fasten one on each side, at the bottom of the opening (figure A).
- Install pivot pins at the top and bottom ends of the door in the prebored holes closest to the jamb (figure B).
- Install doors on pivots.
- Adjust top and bottom pivot sockets in or out so the desired joint is obtained.
- Install pull knobs and door aligners (figure C).

e. Install Pocket Doors

- Install the door frame pocket, when interior partitions are being framed per manufacturer's instructions (figure A).
- Attach rollers to the top of the door.
- Engage the rollers in the track by holding the bottom of the door outward in a way similar to that used with bypass doors.
- Install the stops to the jambs on both sides of the door.

f. Installing Cylindrical Locksets

- Wedge the bottom to hold it in place (figure A).
- Measure up from the floor, usually 36 to 40 inches.
- Position paper template, supplied with the lock, on the striker side edge of the door. Lay out the centers of the holes that need to be bored (figure B).
- Backsets are usually 2 3/8 inches, but 2 3/4 inches is not uncommon. Make sure the backset is marked correctly.
- Bore the hole through the face of the door first; usually 2 1/8 inch in diameter (figure C).
- Score around its faceplate with a sharp pencil. Faceplate markers may be used to lay out the mortise for the latch faceplate (figure D).
- Complete the installation of the lockset by following specific manufacturers' directions.
- The striker plate should be installed so that when the door is closed, it latches snugly with no play (figure E).
- Chisel out the mortise so the plate lies flush with the jamb.
- Screw the plate in place.

Chapter 12 Roofing

Instructor's Note

In this column, you will find a variety of lecture tips to aid student comprehension, notes on when to use PowerPoint slides to teach concepts, safety cautions for your students, and notes on when your class should view a video segment.

Appendix A lists corresponding numbered PowerPoint Presentation titles (see page 197)

Appendix B lists corresponding numbered Video titles and sections as applicable (see page 201)

Introduction

1. Roofing is a general term use to describe materials used to cover a roof and make it weathertight.
2. There are a wide variety of types and styles.
3. Roofs are often installed by specialists.
4. The types discussed here are manufactured with asphalt.
5. Before roofing is applied, the roof deck must be securely fastened. Check that the sheathing is nailed according to the local code.

OBJECTIVES

- Define roofing terms.
- Describe and apply roofing felt underlayment, organic or fiber glass asphalt shingles, and roll roofing.
- Describe and apply flashing to valleys, sidewalls, chimneys, and other roof obstructions.
- Estimate needed roofing materials.

Glossary of Roofing Terms

Apron – the flashing piece that is located on the lower side of a roof penetration such as a chimney or dormer

Asphalt felt – a building paper saturated with asphalt for waterproofing

Blind nail – a method of fastening that conceals the nails

Closed valley – a roof valley in which the roof covering meets in the center of the valley, completely covering the valley

Corner boards – boards used to trim corners on the exterior walls of a building

Cornice – the entire finished assembly where the walls of a structure meet the roof; sometimes called the eaves

Cricket – a small, false roof built behind a chimney or other roof obstacle for the purpose of shedding water around roof penetrations such as a chimney

Drip edge – metal edging strips placed on roof edges to provide a support for the overhang of the roofing material

Eaves – that part of a roof that extends beyond the sidewall

Electrolysis – accelerated oxidation of one metal because of contact with another metal in the presence of water

Exposure – the amount that courses of siding or roofing are exposed to the weather

Flashing – material used at intersections such as roof valleys, dormers, and above windows and doors to prevent the entrance of water

Frieze – a part of the exterior finish applied at the intersection of a overhanging cornice and the wall

Mortar – a mixture of portland cement, lime, sand, and water used to bond masonry units together

Open valley – a roof valley in which the roof covering is kept back from the centerline of the valley

Saddle – same as cricket

Selvage – the unexposed part of roll roofing covered by the course above

Square – the amount of roof covering that will cover 100 square feet of roof area

Water table – finish work applied just above the foundation that projects beyond it and sheds water away from it

A. Asphalt Shingles

1. Provide protection from the weather for a period ranging from twenty to thirty years.
2. Available in many colors and styles.
3. Roofing Terms
   a. A square
      - The amount of roofing required to cover 100 square feet of roof surface.
      - Usually three bundles of shingles per square (figure 12-1).
      - One square of shingles can weight between 235 and 325 pounds.

   b. Deck is the wood roof surface to which roofing materials are applied.
   c. Coverage is the number of overlapping layers of roofing and the degree of weather protection offered by roofing material.
   d. A shingle butt is the bottom exposed edge of a shingle.
   e. Courses are horizontal rows of shingles or roofing.
   f. Exposure is the distance between courses of roofing measured from butt to butt. It is the amount of roofing in each course exposed to the weather (figure 12-2).
   g. The top lap is the height of the shingle or other roofing minus the exposure. In roll roofing this is also known as the selvage.
   h. The head lap is the distance from the butt of an overlapping shingle to the tip of a shingle two courses under measured up the slope.
   i. End lap is the horizontal distance that the ends of roofing in the same course overlap each other.
   j. Flashing are strips of thin sheet metal. They are usually made of lead, zinc, copper, or aluminum.
   k. Asphalt cements and coatings are manufactured and are used as adhesives to bond asphalt roofing products and flashings.
   l. Electrolysis is a reaction that occurs when unlike metals come in contact with water.

2. Roofing Underlayment
   a. Protects the sheathing from moisture until the roofing is applied.
   b. Also gives additional protection after the roofing material is applied.
   c. Typically installed using asphalt felts.

Show examples of various types of shingles to include composition (organic/fiberglass) and various styles (3-tab, laminated, etc.) and colors.

Asphalt roofing products become soft in hot weather. Be careful not to damage them by digging in with heavy shoes during application or by unnecessary walking on the surface after application. Wearing soft rubber-soled shoes causes less wear on the roofing surface during installation.

3. Kinds of Asphalt Shingles
   a. Organic shingles have a base made of heavy asphalt saturated paper felt coated with additional asphalt.
   b. Fiberglass shingles have a base mat of glass fibers.
   c. The granules protect the shingle from the sun, provide color, and protect against fire.
   d. Shingles come in a wide variety of colors, shapes, and weights (figure 12-4).
4. Drip Edge
   a. Is installed along the roof edges.
   b. Usually made of aluminum, galvanized, or painted steel.
   c. Install by using roofing nails of the same metal spaced 8 to 10 inches along its inner edge.
5. Layout of Asphalt Shingles
   a. On smaller roofs, strip shingles are applied by starting from either rake.
   b. On longer buildings, shingles are started at the center and installed both ways.

Snap chalk lines on shop floor to simulate a roof surface. A rectangle at 12′ 3″ × 7′ 2″ will use approximately one square of shingles. Avoid width that won't require cutting ("enders"). Use this area to demonstrate shingle layout. Lines can be moved to simulate crooked ridge lines, etc.

6. Starter Course of Asphalt Shingles
   a. The starter course backs up and fills in the spaces between tabs of the first regular course of shingles.
   b. They have the tabs cut off to apply them with the factory-applied adhesive along the roof perimeter.
7. Fastening Asphalt Shingles
   a. Keep their bottom edges flush with each other. Use a minimum of four fasteners in each strip shingle.
   b. Do not nail into or above the factory-applied adhesive (figure 12-6).
   c. The roofing nail length should be sufficient to penetrate the sheathing at least 3/4 inch, or through approved panel sheathing.
   d. Pneumatic nailers speed up installation significantly (figure 12-7).
8. Shingle Exposure
   a. Most commonly used asphalt shingles have a maximum exposure of 5 inches.
   b. Snap lines across the roof or use the lines of the underlayment, assuming it was installed parallel to the lower edge.
9. Spacing Shingle Courses to the Ridge
   a. The exposure of the last six to ten courses of shingles may be adjusted slightly so the last course has a full exposure where it meets the ridge cap.
   b. Divide this distance into spaces.
   c. Do not exceed the maximum exposure (figure 12-8).
   d. Snap lines across the roof and shingle up to the ridge.

Ridge caps can be cut and nailed to the top of a sawhorse.

10. Ridge Cap
    a. Applied after both sides of the roof have been shingled.
    b. The cap is centered on the ridge.
    c. Exposure for each tab is 5 inches.
    d. Cut shingles. The top edge should be narrower than the bottom edge (figure 12-9).

B. Roll Roofing

1. Roll roofing can be used on roof slopes as little as one inch rise per foot of run.
2. Types of Roll Roofing
   a. Some types are applied with exposed or concealed fasteners.

   b. A concealed-nail type, called double coverage roll roofing, has a top lap of 19 inches.
   c. All kinds come in rolls that are 36 inches wide.
   d. Roll roofing is recommended for use on roofs with slopes less than 4 inches rise per foot.
   e. The exposed-nail type should not be used on pitches less than 2 inches rise per foot.
3. General Application Methods
   a. Cut the roll into 12- to 18-foot lengths.
   b. Spread in a pile on a smooth surface to allow them to flatten out.
   c. Use only the lap or quick-setting cement recommended by the manufacturer.
   d. Apply roll roofing only on a solid, smooth, well-seasoned deck.
   e. Make sure the area below has sufficient ventilation to prevent the deck from absorbing condensation.
   f. A felt underlayment is not used with roll roofing.

C. Flashing

VIDEO #5
Section 3

1. Used to increase the weathertightness at the intersections of different exterior surfaces.
2. It prevents water from entering a building.
3. Flashing material may be sheet copper, zinc, aluminum, galvanized iron, or mineral-surfaced asphalt roll roofing.
4. Eaves Flashing
   a. Whenever there is a possibility of ice dams, install a course of 36-inch wide, self-adhering ice and water shield.
   b. The flashing should extend up the roof far enough to cover at least 12 inches of roof that covers the inside of the wall line of the building (figure 12-11).
5. Open Valley Flashing

Chalk lines can be snapped on shop floor to simulate roofs meeting at an angle to provide for a various valley flashing and shingle layout.

   a. Valleys are finished as open valleys or closed valleys.
   b. Apply a 36-inch wide strip of flashing material centered in the valley.
   c. Lay an 18-inch wide layer of mineral-surfaced roll roofing centered in the valley. Its mineral-surfaced side should be down.
   d. Press the roofing firmly in the center of the valley when nailing the opposite edge.
   e. On top of the first strip, lay a 36-inch wide strip with its surfaced side up.
   f. Fasten it in (figure 12-13).
6. Metal Valleys
   a. Prepare the valley with underlayment in the same manner as described previously.
   b. The metal should extend at least 10 inches on each side of the valley centerline for slopes with a 6-inch rise or less and 7 inches on each side for a steeper slope.
   c. Nail one edge of the flashing with just enough nails to hold it smoothly in place. Use nails of the same metal as the flashing to prevent electrolysis.
   d. Carefully press and form the centerline of the flashing into the valley.
   e. Fasten the remaining edge, using only enough fasteners to hold it (figure 12-14).
7. Closed Valley Flashing
   a. Closed valleys are those where the roof shingles completely cover the valley flashing.
   b. Closed Valley Underlayment

- Apply 30-pound asphalt felt underlayment, centered to the valley.
- Center a 36-inch width of 90-pound smooth or mineral surface roll roofing.

c. Woven Valley Method
- Involves applying asphalt shingles on both sides of the valley and alternately weaving each course over and across the valley (figure 12-15).
- Each course of shingles extends to and over the valley for a distance of at least 12 inches.

d. Closed Cut Valley Method
- Involves applying shingles to one roof area that extends into and past the valley. The other roof is then installed on top, cut to the valley.
- Place fasteners no closer than 6 inches from the center of the valley.

e. Step Flashing Method
- Individual metal pieces tucked between courses of shingles.
- Each piece should be at least 18 inches wide for slopes with a 6-inch rise or greater and 24 inches wide for slopes with less pitch.

8. Flashing Against a Wall

A 4′ × 8′ sheet of plywood can be clamped to a pair of sawhorses to simulate a wall/roof intersection. This can be used in conjunction with lines snapped on the shop floor.

a. The flashings are purchased or cut about 8 inches in width.
b. They are bent at right angles in the center so they will lay about 4 inches on the roof and extend about 4 inches up the sidewall (figure 12-15).
c. Fasten a metal flashing on top of the starter course.
d. Use one fastener in each top corner.
e. Lay the first regular course shingle with its end over the flashing and against the sheathing of the sidewalk.
f. Do not drive any shingle fasteners through the flashings.
g. The remaining step flashings are placed with each course and against the wall.
h. Keep bottom edges about 1/2 inch above the butt of the next course of shingles.

9. Flashing a Chimney

a. A cricket or saddle is built between the upper (back) side of the chimney and roof deck (figure 12-16).
b. Flashings are installed by masons who build the chimney before the carpenter applies the roof covering.
c. The underlayment is applied and tucked under the existing flashings on the sides of the chimney.
d. Along the sides of the chimney, the flashings are tucked in between the shingles in the same manner as in flashing against a wall (figure 12-17).
e. The standing portions of the side flashings are bedded to the chimney with asphalt cement.
f. The head flashing is cemented to the roof. Shingles are applied over it.

10. Flashing Vents

a. Flashings for round pipes usually come as flashing collars made for various roof pitches (figure 12-18).
b. Apply shingle courses under the lower part of the flange.
c. Apply shingles around the stack and over the flange. Do not drive nails through the flashing.
d. Bed shingles to the flashing with asphalt cement, where necessary.

D. Estimating

1. Shingles for a Gable Roof

Take students for a walk/ride and estimate roofing material for various homes.

   a. Multiply ridgeboard length times the rafter length.
   b. Double this to allow for the other side of the roof.
   c. Add to this area three more square feet per lineal foot of ridge for starters and cap shingles.
   d. Divide this sum by 100 to calculate number of squares of shingles.

2. Shingles for a Hip Roof
   a. Multiply building length plus twice the projection times the rafter length.
   b. Double this to allow for the other side of the roof.
   c. Add to this area one more square foot per lineal foot of ridge, eave, and hip.
   d. Divide this sum by 100 to calculate number of squares of shingles.
3. Underlayment
   a. Divide the roof area by the square feet in a roll.

SLIDE 30

E. Procedures

CAUTION Installation of roofing systems involves working on ladders and scaffolding as well as on top of the building. Workers should always be aware of the potential of falling. Keep the location of the roof perimeter in mind at all times.

1. Installing Asphalt Shingles
   a. Clear sawdust and debris that will cause a slipping hazard.
   b. Install underlayment over the deck (figure A).
      - Lay the following courses of felt over the lower course at least 2 inches.
      - Make any end laps at least 4 inches.
      - Lap the felt 6 inches from both sides over all hips and ridges.
      - Nail or staple through each lap and through the center of each layer about 16 inches apart (figure B).
   c. Determine the starting line, either the rake edge or vertical center snapped lines (figure C).
   d. Starting shingle layout at the rake edge involves (figures D and E).
      - Placing the first course, with a whole tab at the rake edge.
      - Starting the second course with a shingle that is 6 inches shorter.
      - Starting the third course with a strip that is a full tab shorter.
      - Starting the fourth with one and a half tabs removed, and so on.
   e. Fasten each shingle from the end nearest the shingle just laid.
   f. Drive fasteners straight so that the nail heads will not cut into the shingles.
   g. Keep the front side edge of the cap shingle to the chalk line.
      - Apply the cap across the ridge until 3 or 4 feet from the end (figure F).
      - Then space the cap to the end in the same manner as spacing the shingle course to the ridge.
      - The last ridge shingle is cut to size.
      - The two fasteners are covered with asphalt cement to prevent leakage.
2. Roll Roofing with Concealed Fasteners
   a. Apply 9-inch wide strips of the roofing along the eaves and rakes (figure A).
   b. Apply the first course of roofing with its edge and ends flush with the strips.
      - Secure the upper edge with nails staggered about 4 inches apart.
      - Do not fasten within 18 inches of the rake edge.
      - Apply cement only to the edge strips covered by the first course.
      - Press the edge and rake edges firmly to the strips.
      - Complete the nails in the upper edge out to the rakes.

c. Apply succeeding courses in like manner.
d. After all courses are in place
- Lift the lower edge of each course.
- Apply the cement in a continuous layer over the full width and length of the lap.
- Press the lower edges of the upper courses firmly into the cement.

3. Covering Hips and Ridges
   a. Cut strips of 12″ × 36″ from a roll of 90-pound rolled roofing.
   b. Bend the pieces lengthwise through their centers.
   c. Snap a chalk line on both sides of the hip or ridge down about 6 inches from the center.
   d. Apply cement between the lines.
   e. Start at the lower end of a hip and at either end of a ridge.
   f. Lap each strip 6 inches over the preceding one.
   g. Nail each strip only on the end that is to be covered by the overlapping pieces.
   h. Spread cement on the end of each strip that is lapped before the next one is applied (figure B).
4. Double Coverage Roll Roofing
   a. Cut the 19-inch strip of selvage, non-mineral surface side, from enough double coverage roll roofing to cover the length of the roof.
   b. Apply the selvage portion parallel to the eaves (figure A).
   c. Apply the first course using a full width strip of roofing. Secure it with two rows of nails in the selvage portion (figure B).
   d. Apply succeeding courses in the same manner.
   e. Lift and roll back the surface portion of each course.
   f. Apply the remaining surfaced portion left from the first course as the last course (figure C).
   g. Specific requirements for quantities and types of adhesive must be followed.
5. Woven Valley Method
   a. Install underlayment and starter strip to both roofs.
   b. Apply first course of one roof into and past the center of the valley.
   c. Apply first course of the other roof in a similar manner, into and past the valley.
   d. Succeeding courses are applied by repeating this alternating pattern (figure A).
6. Closed Cut Valley Method
   a. Begin by shingling first roof completely, letting the end shingle of every course overlap the valley by at least 12 inches.
   b. Snap a chalk line along the center of the valley on top of the shingles of the first roof.
   c. Apply the shingles of second roof, cutting the end shingle of each course to the chalk line.
   d. Place the cut end of each course, that lies in the valley, in a 3-inch-wide bed of asphalt cement (Figure A).

7. Step Flashing Method
   a. Snap a chalk line in the center of the valley on the valley underlayment.
   b. Apply the shingle starter course on both roofs.
   c. Fit and form the first piece of flashing to the valley on top of the starter strips.
   d. Apply the first regular course of shingles to both roofs on each side of the valley, trimming the ends to the chalk line.
   e. Bed the ends in plastic asphalt cement.
   f. Do not drive nails through the metal flashing (figure A).

Chapter 13 Siding and Decks

Instructor's Note

In this column, you will find a variety of lecture tips to aid student comprehension, notes on when to use PowerPoint slides to teach concepts, safety cautions for your students, and notes on when your class should view a video segment.

Appendix A lists corresponding numbered PowerPoint Presentation titles (see page 197)

Appendix B lists corresponding numbered Video titles and sections as applicable (see page 201)

Introduction

1. The portion of the exterior that covers the vertical surface areas of a building is collectively called siding. This does not include masonry coverings, such as stucco or brick veneer.
2. Siding gives the building its architectural look.
3. It must also stand up to the local weather conditions.
4. The area where the lower portion of the roof, or eaves, overhangs the walls is called the cornice. The cornice can be made more or less decorative with variations in detail.
5. Decks are a popular way to increase the living area of a house. Plans may not show specific construction details; therefore, it is important to know the techniques used to build these structures.
6. Scaffolds and ladders are used during this stage of construction. Every safety concern previously stated applies here.

OBJECTIVES

- Describe the shapes, sizes, and materials used as siding products.
- Install corner boards and prepare side wall for siding.
- Apply horizontal and vertical siding.
- Apply plywood and lapped siding.
- Apply wood shingles and shakes to sidewalls.
- Apply vinyl and aluminum siding.
- Describe various types of cornices and name their parts.
- Install gutters and downspouts.
- Describe the construction of and kinds of materials used in decks.
- Lay out and install footings, supporting posts, girders, and joists for a deck.
- Apply decking in the recommended manner and install flashing, for an exposed deck, against a wall.

Glossary of Siding and Deck Terms

Baluster – vertical members of a stair rail, usually decorative and spaced closely together

Battens – a thin, narrow strip usually used to cover joints in vertical boards

Blind nail – a method of fastening that conceals the fastener

Corner boards – finish trim members used at the intersection of exterior walls

Cornice – a general term used to describe the part of the exterior finish where the walls meet the roof

Downspout – a vertical member used to carry water from the gutter downward to the ground; also called leader

Drip – that part an exterior finish that projects below another to cause water to drop off instead of running back against and down the wall

Eaves – the part of the roof that extends beyond the sidewalls

Frieze – that part of exterior trim applied to cover the joint between the overhanging cornice and the siding

Gutter – a trough attached to an eave used to carry off water

Plancier – the finish member on the underside of a box cornice; also called soffit

Soffit – the horizontal, underside trim member of a cornice or other overhanging assembly

Striated – finish material with random and finely spaced grooves running with the grain

Story Pole – a narrow strip of wood used to lay out the heights of courses of siding or members of a wall frame.

Water table – exterior trim members applied at the intersection of the siding and the foundation that projects outward to direct water away from the building

A. Siding

1. Types of Sidings

Take students for a field trip to identify various sidings and cornices. Note all accessories and trim details.

   a. Siding is manufactured from solid lumber, plywood, hardboard, aluminum, cement, and vinyl.
   b. Siding may be applied horizontally, vertically, or in other direction, to make many interesting designs (figures 13-1 and 13-2).
   c. Bevel siding is one widely used variety (figure 13-3).
   d. Panel siding
      - Made from plywood.
      - Comes in several thicknesses, 4-foot widths, and lengths of 8, 9, and 10 feet. It is usually applied vertically.
      - Most panel siding is shaped with shiplapped edges for weathertight joints.
   e. Lap siding
      - Manufactured with a variety of materials and surface textures.
      - They come in thicknesses from 7/16 to 13/16 inch, widths from 4 to 12 inches, and lengths of up to 12 feet (figure 13-5).
2. Preparing for Siding Application
   a. Foundation Trim
      - The siding is started so that it extends slightly below the top of the foundation.
      - A water table may be installed for appearance.
   b. Eaves Treatment
      - Siding may end against the bottom edge of the frieze.
      - Siding may also terminate against the soffit.
   c. Rake Trim
      - Siding may be applied under a furred-out rake fascia.
      - When the rake overhangs the sidewall, the siding may be fitted against the rake soffit. When fitted against the rake soffit the joint is covered with a molding (figure 13-6).

d. Treating Corners
- Corner boards.
- Horizontal siding may also be mitered around exterior corners.
- Metal corners may be used on each course of siding.
- Interior corners, siding courses may butt against a square corner board or against each other (figure 13-7).

e. Applying Building Paper
- The housewrap should lap over any flashing applied at the sides and tops of windows and doors at corner boards.
- It should be tucked under any flashings applied under the bottoms of windows or frieze boards.

3. Installing Horizontal Siding
   a. Exposure of courses of bevel siding can be varied somewhat.
   b. Once the siding is laid out for the desired exposure the layout is transferred to other locations with a story pole.
4. Installing Vertical Tongue and Groove Siding
   a. Corner boards usually are not used when wood siding is applied vertically (figure 13-9).
   b. The starting piece has the grooved edge removed. The tongue edge should be plumb.
   c. The bottom end should be about 1 inch below the sheathing.
   d. The top end should neatly butt or be tucked under any trim above.
   e. Blind nail through the tongue edge.
   f. Nails should be placed from 16 to 24 inches apart.
   g. Blocking must be provided between studs if siding is applied directly to the frame.
5. Installing Panel Siding
   a. Panel siding may installed both vertically and horizontally.
   b. It may be applied to sheathing or directly to studs if backing is provided for all joints (figure 13-10).
   c. All vertical edges should start and end on a stud. Fasten panels of 1/2-inch thickness.
   d. Leave a 1/8-inch space between panels.
6. Wood Shingle and Shake Siding
   a. Shingle and shake siding come in a variety of shapes and sizes (figure 13-11).
   b. Special fancy butt shingles provide interesting patterns, in combination with square butts or other types of siding (figure 13-12).
   c. In single coursing, shingles are applied to walls with a single layer in each course.
   d. In Double Coursing
      - Two layers are applied in one course.
      - Even greater weather exposures are allowed.
      - Used when wide courses with deep, bold shadow lines are desired (figure 13-13).
   e. Corners
      - Shingles may be butted to corner boards like any horizontal wood siding.
      - May also be applied by alternately overlapping each course in the same manner as in applying a wood shingle ridge.

- Inside corners may be woven by alternating the corner shingle first on one wall and then the other (figure 13-14).

7. Vinyl and Aluminum Siding
   a. Siding Panels
      - For horizontal application, are made in 8-, 10-, and 12-inch widths.
      - Come in configurations to simulate one, two, or three courses of bevel or drop siding.
      - Panels designed for vertical application come in 12-inch widths. They are shaped to resemble boards.
      - Can be used in combination with horizontal siding.
      - For ventilation, perforated soffit panels of the same configuration are used (figure 13-16).

   b. Siding System Accessories
      - Inside and outside corner posts.
      - J-channels (figure 13-17)
      - The majority of vinyl siding panels and accessories are manufactured in lengths of 12 feet.

   c. Applying Horizontal Siding
      - Center fasteners in the nailing slots.
      - Do not drive them too tightly (figure 13-18)
      - Space fasteners 16 inches apart for horizontal siding and 6 to 12 inches apart for accessories unless otherwise specified by the manufacturer.
      - Install J-channel across the top and along the sides of window and door casings. It may also be installed under windows or doors with the undersill nailed inside the channel.
      - To cut, use tin snips, hacksaw, utility knife, or power saw with an abrasive wheel or fine-tooth circular blade. Reverse the blade, if a power saw is used, for smooth cutting.

   d. Gable End Installation
      - Trim the rakes of a gable end first with J-channels.
      - Make a pattern for cutting gable end panels at an angle where they intersect with the rake (figure 13-19).

   e. Applying Vertical Siding
      - Space fasteners about 12 inches apart for vertical siding panels.
      - The starter strip is the narrower 1/2-inch J-channel or drip cap flush with and fitted into the corner posts (figure 13-20).

   f. Brake
      - Aluminum coil stock is sold in rolls of various widths from 12 to 24 inches and lengths of 50 feet.
      - Coil stock may be cut by using a utility knife, then bent to the desired configuration.
      - Bending the stock begins with careful positioning of the piece in the jaws (figure 13-21).
      - Lock the jaws tightly to secure the piece during the bend (figure 13-22).
      - A full 180-degree bend takes two steps. First, align the piece to as small a bend as possible, usually about 5/8 inches. Lock and bend as far as the brake will allow. Unlock jaws, remove piece, and relock jaws, placing piece on top of the jaws (figure 13-23). The final bend to 180 degrees is made on top.

B. Cornices

1. Cornice is a term used to describe the trim and accessories used to finish the joint between the siding and the roof. It has many parts and shapes.
   a. Subfascia
      - Sometimes called the false fascia or rough fascia (figure 13-24).
      - Fastened to the rafter tails.
      - Generally a nominal 1- or 2-inch-thick piece.

   b. Soffit
      - The underside of the cornice is the plancier, often referred to as a soffit (figure 13-25).
      - Material may include solid lumber, plywood, strand board, fiberboard, or corrugated aluminum and vinyl panels.
      - Soffits should be ventilated.

VIDEO #4
Section 9

   c. Fascia
      - Is fastened to the subfascia, or to the ends of the rafter tails when the subfascia is not used.
      - May be a piece of lumber grooved to receive the soffit.
      - Also may be made from bent aluminum and vinyl used to wrap the subfascia.
      - The portion of the fascia that extends below the soffit is called the drip.

   d. Frieze
      - Fastened to the sidewall with its top edge against the soffit.
      - Its bottom edge is rabbeted or furred to receive the sidewall finish.

   e. Cornice Molding
      - Used to cover the joint between the frieze and the soffit.

   f. Lookouts
2. Cornice Design
   a. Cornices are generally classified into three main types: box, snub, and open (figure 13-26).
      - The Box Cornice
        1) Probably the most common.
        2) Gives a finished appearance to this section of the exterior.
      - The Snub Cornice
        1) Has no rafter overhang.
      - The Open Cornice
        1) Has no soffit.
        2) A contemporary or rustic design.

   b. Rake Cornices
      - The cornices of eave and rake must be blended together to make a smooth transition (figure 13-27).

   c. Cornice Returns (Figure 13-28)

C. Gutters and Downspouts

1. Miscellaneous Facts
   a. Gutters and downspouts carry water down the wall and away from the foundation (figure 13-29).
   b. Gutters are usually made of aluminum, copper, or vinyl (figure 13-30).

c. Gutters are installed to slope in one direction with a downspout at the end.

d. Gutter systems have components comprised of

- Inside corners.
- Outside corners.
- Joint connectors.
- Outlets.
- Caps.
- Brackets (figure 13-31).

D. Decks

1. Wood decks and porches are built to provide outdoor living areas (figure 13-32).
2. Decking materials are chosen for strength and durability, as well as appearance and resistance to decay.
3. Lumber
   a. Pressure-treated southern pine.
   b. Western cedar.
   c. California redwood.
   d. #2 grade is structurally adequate for most applications.
4. Lumber Sizes
   a. Depend on the spacing and height of supporting posts and the spacing of girders and joists.
   b. Depend on the type of wood used and the weight imposed on the members.
5. Fasteners
   a. Stainless steel.
   b. Hot-dipped galvanized.
6. Building a Deck
   a. Posts, set on footings.
   b. Supporting a platform.
   c. Girders and joists.
   d. Deck boards.
   e. Post, rails, balusters, and other special parts make up the railing (figure 13-33).
   f. A ledger is nailed or bolted to the wall and acts as a beam to support joists.
   g. Footings
      - Under each post.
      - Usually made twice the width of the post it is supporting.
      - In cold climates the footing should extend below the frost line.
      - Posts must be firmly attached to the footings
7. Stairs
   a. To protect the bottom ends of the stair carriage, they should be treated with preservative and supported by an above-grade concrete pad (figure 13-34).
   b. Stairs with more than two risers are generally required to have at least one handrail (figure 13-35).
8. Rails
   a. There are numerous designs for deck railings (figures 13-36 and 13-37).
   b. All designs must conform to certain code requirements.

c. Most codes require at least a 36-inch high railing around the exposed sides, if the deck is more than 30 inches above the ground.
d. No openings in the railing should allow a 4-inch sphere to pass through.

9. Deck Accessories
   a. Shading structures.
   b. Benches.

E. Estimating

Gain permission from a homeowner or use buildings on campus as practice examples for estimating sidings.

1. Siding
   a. Calculate the wall area.
   b. Add the gable area and dormers, bays, and porches to get the total building area.
   c. Deduct the total window and door area (figure 13-38).
   d. Add a waste factor for cuts as needed.
2. Panel Siding
   a. Building perimeter multiplied times wall height.
   b. Then divide by the area per sheet.
3. Shingle Siding
   a. The number of squares of shingles needed to cover a certain area depends on how much they are exposed to the weather.
   b. Divide the wall area by the coverage of a bundle.
4. Aluminum or Vinyl Siding
   a. Sold by the square.
   b. Determine the wall area.
   c. Deduct for the openings.
   d. Divide by 100.
   e. Add for waste as needed.
   f. Measure the total linear feet required for each accessory.

F. Procedures

1. Installing Horizontal Siding
   a. First determine the siding exposure (figure A).
   b. Install a starter strip (figure B).
   c. Layout the desired exposures on each corner board and each side of all openings.
   d. Snap lines at these layout marks.
   e. Install the siding as per manufacturer's recommendations (figures C and D).
   f. Siding is fastened to each bearing stud.

SLIDE 31

2. Installing Vertical Tongue and Groove Siding
   a. Place first piece vertically on the wall with the beveled edge flush with the corner. Face nail the edge nearest the corner (figure A).
   b. Fasten a temporary piece on the other end and stretch a line to keep the bottom ends of other pieces in a straight line.
   c. Apply succeeding pieces by toenailing into the tongue edge of each piece.
   d. If joints are necessary, use a mitered or rabbeted joint for weathertightness (figure B).
   e. Use a scrap block to mark the first cut piece around an opening (figure C).
   f. Repeat previous marking concept to complete the cut piece around opening (figure D).

3. Installing Panel Siding
   a. Install the first sheet with its long edge plumb.
   b. The factory edge should fall on the center of a stud. Panels must also be installed with their bottom ends in a straight line.
   c. Horizontal butt joints must be either offset and lapped, rabbeted, or flashed (figure A).
4. Installing Wood Shingles and Shakes
   a. Fasten a shingle on both ends of the wall with its butt about 1 inch below the top of the foundation.
   b. Stretch a line between them (figure A).
   c. Fill in the remaining shingles to complete the undercourse.
   d. Apply another course on top of the first course.
   e. Offset the joints in the outer layer at least 1 1/2 inches
   f. Shingles should be spaced 1/8 to 1/4 inch apart to allow for swelling (figure B).
   g. To apply the second course, snap a chalk line across the wall and tack 1 × 3 straightedges to the wall with their top edges to the line (figure C).
5. Applying Horizontal Vinyl Siding
   a. Snap a level line to the height of the starter strip.
   b. Fasten the strips to the wall (figure A).
   c. Cut the corner posts so they extend 1/4 inch below the starting strip.
   d. Cut each J-channel.
   e. Snap the bottom of the first panel into the starter strip.
   f. Leave a 1/4 inch space in the corner post channel (figure B).
   g. Overlap each panel about 1 inch.
   h. The exposed ends should face the direction from which they are least viewed.
   i. To fit around a window, install the panel under the window and up in the undersill trim (figures C and D).
   j. Use a specialized snaplock punch (figure E).
   k. Panels are cut and fit over windows (figure F).
   l. Install the last course of siding panels under the soffit in a manner similar to fitting under a window (figure G).
6. Applying Vertical Vinyl Siding
   a. Measure and lay out to determine the width of the first and last piece.
   b. Cut the edge of the first panel nearest the corner.
   c. Install an undersill trim in the corner board or J-channel.
   d. Punch lugs along the cut edge.
   e. Snap the panel into the undersill trim (figure A).
   f. Install the remaining full strips, making sure there is 1/4-inch gap at the top and bottom.
7. Installing Gutters (figure A)
   a. On both ends of the fascia, mark the location of the bottom side of the gutter.
   b. Stretch a chalk line between the two marks.
   c. Fasten the gutter brackets to the chalk line on the fascia with screws.
   d. Locate and install the outlet tubes in gutter as required.
   e. Hang the gutter sections in the brackets.
   f. Fasten downspouts to the wall with appropriate hangers and straps (figures B and C).

8. Building a Deck
   a. Install the ledger (figure A).
   b. Place all supporting posts on the footings (figures B, C, and D).
   c. Install the girders on the posts (figure E).
   d. Lay out and install the joists (figure F).
   e. If the deck is 4 feet or more above the ground, the supporting posts should be braced (figure G).
   f. Flashing is installed under the siding and on top of the deck.
   g. Lay boards with bark side up or annular rings arching downward to minimize cupping (figure H).
      - Start at the outside edge if the deck is built against a building.
      - Maintain about a 1/4-inch space between boards.
      - If the decking boards are green or wet, nail them tight together.
   h. Cut the boards so their butt ends are centered over joists (figure I).
   i. Stagger them between adjacent rows.
   j. Predrill holes for fasteners at the board ends to prevent splitting.

Chapter 14 Insulations and Wall Finish

Instructor's Note

In this column, you will find a variety of lecture tips to aid student comprehension, notes on when to use PowerPoint slides to teach concepts, safety cautions for your students, and notes on when your class should view a video segment.

Appendix A lists corresponding numbered PowerPoint Presentation titles (see page 197)

Appendix B lists corresponding numbered Video titles and sections as applicable (see page 201)

Introduction

1. At this point in the construction process it is assumed the initial stage, or "rough in" of the mechanicals (plumbing, heating, cooling, electrical components) has been completed.
2. The carpenters are set to return and begin the interior finish.
3. Insulation is a material installed to resist the transfer of heat.
4. It is important for the carpenter to understand how insulation works, what its workability requirements are, and how to get the most out of insulation.
5. Excess moisture causes framing to rot or corrode and insulation to perform badly. Careful attention to details will keep most moisture out, and ventilation will remove the rest.
6. The majority of interior wall and ceiling surfaces in homes today are made with gypsum board.

OBJECTIVES

- Describe how insulation works and define insulating terms and requirements.
- Describe the commonly used insulating materials and state where insulation is placed.
- Properly install various kinds of insulation.
- State the purpose of and install vapor retarders.
- Explain the need for ventilating a structure and describe types of ventilators.
- Describe various kinds, sizes, and uses of gypsum panels.
- Describe the kinds and sizes of nails, screws, and adhesives used to attach gypsum panels.
- Apply gypsum board to interior walls and ceilings.
- Conceal gypsum board fasteners and corner beads.
- Reinforce and conceal joints with tape and compound.
- Describe and apply several kinds of sheet wall paneling.
- Describe and apply various patterns of solid lumber wall paneling.
- Estimate quantities of drywall, drywall accessories, and sheet and board wall paneling.

Glossary of Insulation and Wall Finish Terms

Air infiltration – unwanted movement of air into an insulation layer or a conditioned space (heated or cooled)

Condensation – when water, in a vapor form, changes to a liquid due to cooling of the air; the resulting drops of water that accumulate on the cool surface

Deadman – a T-shaped wood device used to support ceiling drywall panels and other objects

Dew point – temperature at which moisture begins to condense out of the air

Eased edge – an edge of lumber whose sharp corners have been rounded

Eaves – that part of a roof that extends beyond the sidewall

Face – the best-appearing side of a piece of wood or the side that is exposed when installed

Gypsum board – a panel used as a finished surface material made from a mineral mined from the earth

Insulation – material used to restrict the passage of heat or sound

Hardboard – a building product made by compressing wood fibers into sheet form

R-value – a number given to a material to indicate its resistance to the passage of heat

Storm sash – an additional sash placed on the outside of a window to create dead air space to prevent the loss of heat from the interior in cold weather

Vapor retarder – a material used to prevent the passage of vapor

Wainscoting – a wall finish applied partway up the wall from the floor

Weatherstripping – narrow strips of thin metal or other material applied to windows and doors to prevent the infiltration of air and moisture

A. Insulation and Ventilation

1. How Insulation Works
   a. All materials used in construction have some insulating value, some better than others.
   b. R-value indicates the materials' ability to resist heat transfer.
   c. The higher the number the better the insulator (figure 14-1).
   d. Materials Used for Insulating
      - Glass fibers.
      - Mineral fibers (rock).
      - Organic fibers (paper).
      - Plastic foam.
      - Aluminum foil is also used to reflect heat radiation.
2. Insulation Requirements
   a. Average winter low-temperature zones of the United States are shown in figure 14-2.
   b. This information is used to determine the R-value of insulation installed in walls, ceilings, and floors (figure 14-3).
   c. In warmer climates air-conditioned homes receive insulation in walls, ceilings, and floors.
3. Where Heat is Lost (figure 14-4)
   a. Windows and doors are generally sources of great heat loss.
   b. Heat loss through glass surfaces can be reduced 50 percent or more by installing double- or triple-glazed windows or by adding a storm sash and storm doors.
   c. Weatherstripping around windows and doors also reduces heat loss.
4. Ventilation
   a. An air space must be maintained between the insulation and the roof sheathing (figure 14-5).

B. Types of Insulation

1. Flexible Insulation
   a. Blanket Insulation

Have examples of insulation types, vapor retarders, and vents.

- Comes in rolls (figure 14-6).
- Widths are suited to 16- and 24-inch stud and joist spacing.
- Thicknesses up to 12 inches.
- Made of fluffy rock or glass wool fibers.
- May be purchased with a facing of asphalt-laminated draft paper or aluminum foil with flanges on both edges for fastening to studs or joists.
- Facing of the blanket serves as a vapor retarder.
- Facing should always face the warm side of the wall.

   b. Batt Insulation (figure 14-7)

- Made of the same material and facings as blanket insulation.
- Comes in the same thicknesses and widths.
- Lengths are either 48 or 93 inches.

2. Loose-Fill Insulation
   a. It is supplied in bags or bales.
   b. It is placed by pouring, blowing, or packing by hand.
   c. Materials include rock or glass wool, wood fibers, shredded redwood bark, cork, and wood pulp products.
   d. Well suited for use between trusses and ceiling joists.
3. Rigid Insulation
   a. Available in widths of 2 or 4 feet and lengths of 8 feet.
   b. Polystyrene comes in two forms, expanded and extruded.

CAUTION Aluminum foil conducts electricity very well. Care should be employed when installing reflective insulation around electrical wiring.

- Expanded polystyrene is a white bead board similar to foam coffee cups.
- Extruded polystyrene is the only insulation board that may come in contact with water and still be an effective insulator.

   c. Polyurethane foams

- Have the best R-value per inch.
- Usually foil faced and sometimes faced with Kraft and asphalt-impregnated paper.

   d. Rigid insulation is easily cut with a knife or (table) saw.
   e. It is recommended that respirators be worn when using an electric saw to cut rigid foam panels.
4. Reflective Insulation
   a. Consists of outer layers of aluminum foil bonded to inner layers of various materials.
   b. Installed facing an air space with a depth of 3/4 inch or more.
5. Foamed-in-place insulation
   a. This product is installed by professionals using highly specialized tools and equipment.
   b. Foamed-in-place insulation performs the best of all insulation materials.

C. Installing Insulation

1. General Principles
   a. Pay attention to details around outlets, pipes, and any obstructions.
   b. Keep the natural fluffiness of the insulation at all times.

   c. Polyethylene should be placed on the soil of crawl spaces.
   d. Provision should also be made for ventilation of unheated areas.
2. Installing Flexible Insulation
   a. Insulation that is not properly installed can drastically reduce the effectiveness of the insulation.
   b. For step-by-step instructions on installing flexible insulation, see the Procedures section.

D. Condensation

1. As moisture-laden air is cooled, it reaches the dew point.
2. The natural process is detrimental to buildings.
3. Moisture in Buildings
   a. It is produced mainly by cooking, taking showers and baths, washing and drying clothes, and using a humidifier.
   b. Water vapor condenses under certain conditions when it comes in contact with a cooler surface (figure 14-13).
   c. Ice dams
      - Heat from the attic can cause the snow next to the roof to melt.
      - This causes the running water to back up on the roof, under the shingles, and into the walls and ceilings (figure 14-14).
4. Problems Caused by Condensation
   a. If insulation absorbs water, the thermal resistance and effectiveness of the insulation is reduced.
   b. Bacteria and fungi grow.
   c. Wood decays.
   d. The external paint can blister and peel.
5. Prevention of Vapor Condensation
   a. Reduce the amount of moisture within the house.
   b. Reduce the moisture migration into the insulation layer.
      - Airtight construction techniques are the best way to reduce moisture migration into the insulation layer.
      - Installing a vapor retarder is the common method to create this air seal.

E. Ventilation

1. General Principles
   a. Helps combat condensation of moisture.
   b. Attic
      - Melting snow on the roof over the attic space can be eliminated and ice damming is eliminated.
      - Roof shingles will stay cooler in the summer months and thus last longer.
   c. Crawl spaces.
   d. On roofs where the ceiling finish is attached to the roof rafters.
2. Ventilating Gable Roofs
   a. Triangularly-shaped louver vents are often installed in both end walls of a gable roof.
   b. Continuous ridge and soft vents (figure 14-17).
   c. Positive-ventilation chutes should be installed to prevent any air obstructions by the ceiling insulation near the eaves.

3. Ventilating Hip Roofs
   a. Continuous venting along each hip rafter (figure 14-18).
4. Ventilating Crawl Spaces
   a. Rectangular metal or plastic louvered ventilators are used (figure 14-19).
   b. Ground cover vapor retarder is highly recommended.

F. Drywall Construction

1. Gypsum Board
   a. Fire-resistant, readily available, easy to apply, decorate, or repair, and is relatively inexpensive.
2. Types of Gypsum Panels
   a. An aluminum foil backing functions as a vapor retarder.
   b. Regular gypsum panels are most commonly used.
   c. An eased-edge gypsum board has a special tapered, rounded edge.
   d. Type X gypsum board is commonly known as firecode board or X-rock.
   e. Moisture resistant (MR) gypsum board, or green board, consists of special moisture-resistant core and paper cover that is chemically treated to repel moisture.
   f. Special Purpose
      - Backing board is designed to be used as a base layer in multilayer systems.
      - Core-board is available in 1-inch thickness and used for the core of solid gypsum partitions.
      - Predecorated panels have coated, printed, or overlay surfaces that require no further treatment.
      - Liner board that has a special fire-resistant core is used to cover shaft walls, stairwells, chaseways, and similar areas.
   g. Veneer plaster base is commonly called blue board and is designed to receive applications of veneer plaster.
3. Sizes of Gypsum Panels
   a. Panels are manufactured 4 feet wide and in lengths of 8, 9, 10, 12, 14, or 16 feet.
   b. Thicknesses
      - A 1/4-inch thickness is used as a base layer in multilayer applications.
      - A 3/8-inch thickness is usually applied as a face layer in repair and remodeling work over existing surfaces.
      - Both 1/2 inch and 5/8 inch are commonly used thicknesses of gypsum panels for single-layer wall and ceiling application in residential and commercial construction.
4. Cement Board
   a. Has a core of portland cement reinforced with a glass fiber mesh embedded in both sides (figure 14-23).
   b. Will not deteriorate when wet.
   c. Used extensively around bathtub, shower, kitchen, and laundry areas as a base for ceramic tile.
   d. Standard cement board panels come in a thickness of 1/2 inch and 5/8 inch.
   e. Cement board is also manufactured in a 5/16 inch thickness. It is used as an underlayment for floors and countertops.
5. Drywall Fasteners
   a. Nails should be driven with a drywall hammer designed to compress the gypsum panel face to form a dimple (figure 14-25).

CAUTION Drywall fasteners often have metal shavings still attached to the threads. These shavings can become metal splinters if, while holding the screw to start, it is spun between the thumb and finger.

b. Drywall screws are used to fasten gypsum panels.
   - Driven with a drywall screwgun (figure 14-26).
   - Driven in the panel surface without breaking the paper (figure 14-27).
   - Driven so the fastener head can later be covered with compound.

c. Adhesives
   - Used to bond single layers directly to supports.
   - Laminate gypsum board to base layers.
   - Supplemental fasteners must be used with stud adhesives (figure 14-28).

6. Techniques for Drywall Fitting
   a. Cutting and Fitting Gypsum Board
      - For step-by-step instructions on cutting and fitting gypsum board, see the Procedures section.
   b. Making Cutouts in Wall Panels
      - Measuring and cutting.
      - Electric drywall cutout tool.
   c. Drywall is fastened to framing members with nails or screws.
   d. Single Nailing Method – nails are spaced a maximum of 7 inches OC on ceilings and 8 inches OC on walls into frame members.
   e. Double Nailing Method – 12 inches OC with a second set about 2 inches from the first set (figure 14-32).
   f. Screw Attachment – spaced 12 inches OC on ceilings and 16 OC on walls.
   g. Using Adhesives.
7. Installing Drywall
   a. Ceiling
      - Gypsum panels are applied first to ceilings and then to the walls.
      - Panels may be applied parallel, or at right angles, to joists or furring.
      - For step-by-step instructions on installing ceiling drywall, see the Procedures section.
   b. Installing Horizontal Drywall
      - Stagger end joints.
      - For step-by-step instructions on installing horizontal drywall, see the Procedures section.
   c. Vertical Application on Walls
      - All cut edges must be in the corners.
      - Tapered edge should be plumb and centered on the stud (figure 14-36).
   d. Floating Interior Angle Construction
      - Fasteners are omitted in the corner intersection of the ceiling and wall panels.
      - Panels applied on walls are fitted tightly against ceiling panels.

G. Concealing Fasteners and Joints

1. Description of Materials
   a. Joint Compounds
      - An all-purpose compound is used for both bedding the tape and finishing the joint.
      - Setting type joint compounds are used when a faster setting time is desired.
   b. Joint Reinforcing Tape
      - Used to cover, strengthen, and provide crack resistance to drywall joints.

- High-strength fiber paper is designed for use with compounds on gypsum panels.
- Glass fiber mesh is designed to reinforce joints on veneer plaster gypsum panels (figure 14-40).

c. Corner Bead and Other Drywall Trim

- Corner beads are applied to protect exterior corners of drywall construction (figure 14-41).
- Metal corner tape is used for corner protection on arches, windows with no trim, and other locations (figure 14-43).
- Metal trim provides protection and finished edges to drywall panels (figure 14-44).
- Control joints are metal strips with flanges on both sides of a 1/4-inch, V-shaped slot (figure 14-45).

2. Applying Joint Compound and Tape

a. Maintain the interior temperature at a minimum of 50 degrees Fahrenheit.
b. Avoid contamination of the compound.
c. A taping tool is sometimes used (figure 14-47).
d. For step-by-step instructions on applying joint compound and tape see the Procedure sections.

H. Sheet Paneling

1. Plywood

a. Prefinished plywood is probably the most widely used sheet paneling.
b. Available in both hardwoods and softwoods.
c. Less expensive kinds are prefinished with a printed wood grain or other design on a thin vinyl covering.
d. Panel thicknesses are 3/16 and 1/4 inch.
e. Four feet wide and 7 to 10 feet long (figure 14-48).
f. Matching molding is available to cover edges, corners, and joints.
g. Thin ring-shanked nails, called color pins, are used when exposed fastening is necessary.

2. Hardboard

a. Designed simulate stone, brick, stucco, leather, weathered or smooth wood, and other materials.
b. Four feet wide and in lengths of from 8 to 12 feet.
c. Thicknesses are from 1/8 to 1/4 inch.

3. Particleboard

a. Sheets are usually 1/4-inch thick, 4 feet wide, and 8 feet long.
b. Sheets are brittle and break easily and must be applied only on a solid wall backing.

4. Board Paneling

a. Wood Species

- Light tones include birch, maple, spruce, and white pine.
- Medium tones are cherry, cypress, hemlock, oak, ponderosa pine, and fir.
- Darker-toned woods are cedar, mahogany, redwood, and walnut.

b. Surface Textures and Patterns

- Planed or rough-sawn.

- Square-edge.
- Tongue-and-grooved or shiplapped (figure 14-49).

c. Sizes

- Most wood paneling comes in a 3/4-inch thicknesses and in nominal widths of 4, 6, 8, 10, and 12 inches

SLIDE 31

5. Installation of Sheet Paneling
   a. A baker board layer, at least 3/8-inch thick, should be installed on walls prior to the application of sheet paneling (figure 14-51).
   b. Partial paneling is called wainscoting and usually installed about 32 inches above the floor (figure 14-52).
   c. Installation begins in a corner.
   d. To fasten with nails only, space them about 6 inches apart along edges and about 12 inches apart on intermediate studs for 1/4-inch thick paneling (figure 14-53).
   e. If adhesives are used, apply a 1/8-inch continuous bead where panel edges and ends make contact.
6. Installing Solid Wood Board Paneling
   a. The boards are fastened to the existing wall studs.
   b. For vertical application of board paneling in a frame wall, blocking must be provided between studs.
   c. Allow the boards to adjust to room temperature and humidity.
   d. If boards are a uniform width, the width of the starting board must be planned to avoid ending with a narrow strip.

I. Estimating

Measure room of school building and estimate materials needed for drywall.

1. Drywall
   a. To find the ceiling area, multiply the length of the room by its width.
   b. To find the wall area, multiply the perimeter of the room by the height.
   c. Subtract all large wall openings.
   d. Add about 5 percent of the total for waste.
2. Drywall Accessories
   a. About 1,000 screws are needed for each 1,000 square feet of drywall when applied to framing 16 inches OC.
   b. 850 screws are needed for 24-inch OC framing.
   c. Approximately 370 feet of joint tape and 135 pounds of conventional, ready-mixed joint compound are needed for every 1,000 square feet.
3. Sheet Paneling
   a. Divide the perimeter by the width of the panels to be used.
   b. Deduct the area of any large openings.
   c. Multiply the area to be covered by the area factor shown in figure 14-56.

J. Procedures

1. Installing Flexible Insulation
   a. Install positive ventilation chutes between the rafters (figure A).
   b. Install the air-insulation dam between rafters.
   c. Using another scrap piece of wood, compress the insulation, and cut it with a sharp knife in one pass (figure B).
   d. Place the batts or blankets between the studs (figure C).

e. Fill any spaces around windows and doors (figure D).

f. Install ceiling insulation.

g. Flexible insulation installed between floor joists over crawl spaces may be held in place by wire mesh or pieces of heavy-gauge wire wedged between the joists (figure E).

2. Cutting and Fitting Gypsum Board

   a. Take measurements accurately to within 1/4 inch for the ceiling and 1/8 inch for the walls.

   b. Cut the board by first scoring the face side through the paper. Guide it with a drywall T-square (figure A).

   c. Bend the board back against the cut. Score the backside paper (figure B).

   d. To make cuts parallel to the long edges, the board is often gauged with a tape and scored with a utility knife (figure C).

   e. Ragged edges can be smoothed with a drywall rasp.

3. Making Cutouts in Wall Panels

   a. Cutout Tool Method
      - Note the location of the center of the outlet box.
      - Install the wall panel over the box with only enough fasteners to hold the panel in place.
      - Using the cutout tool, plunge a hole through the panel.
      - Move the tool until the bit touches a side of the box.
      - Withdraw and reinsert the tool and trace the outside of the box (figure A).

   b. Layout and Cut Method
      - Measure the height from the floor, ceiling, or edge of previous sheet.
      - Measure the outline of the box to be cut on the panel.
      - With a saw or utility knife, cut the outline of the box (figure A).

4. Installing a Ceiling Drywall

   a. Cut the first board to width and length to within 1/4 inch.

   b. Lay out lines on the panel face indicating the location of the framing in order to place fasteners accurately.

   c. Lift the panel overhead and place it in position (figure A).

   d. Fasten the panel using the appropriate fastener spacing (figure B).

5. Installing Horizontal Drywall

   a. Cut the panel to length about 1/8 short.

   b. Raise the top panel into place and hold firmly against the ceiling (figure A).

   c. Drive the fasteners and finish the fastening.

   d. Measure and cut the bottom panel.

   e. Raise the panel into position, holding firmly against upper sheet.

   f. Stagger any necessary end joints.

   g. Where end joints occur on steel studs, attach the end of the first panel to the open or unsupported edge of the stud (figure B).

6. Applying Joint Compound and Tape

   a. Fill the recess formed by the tapered edges of the sheets with joint compound. Use a joint knife (figure A).

   b. Center the tape on the joint, embed the tape, and remove excess compound (figure B).

Screw a sheet of gypsum board to a sheet of plywood. Mount this assembly to wall and ceiling and have students practice installing screws. Draw parallel lines 1″ apart and have students drive 100–200 screws in wall and 100–200 screws in ceiling. If screws are driven 10″ apart, each sheet should accept approximately 4,000 screws.

If new gypsum board is covered with three coats of water-borne urethane, joint compound can be applied and removed while still wet, so that the board may be used several times to facilitate training.

   c. Spot to conceal fastener heads.
   d. The first coat of compound is applied about 6 inches wide from the nose of the bead to a feather edge on the wall.
   e. Allow the first coat to dry thoroughly.
   f. The second coat is feathered out about 2 inches beyond the edges of all first coats, approximately 7 to 10 inches wide.
   g. Apply a third coat of compound over all fill coats.
   h. At interior corners the tape is folded in the center to fit in the corner.
7. Installing Sheet Paneling
   a. Starting the Application
      - Mark the location of each stud in the wall on the floor and ceiling.
      - Apply narrow strips of paint on the wall from floor to ceiling over the stud where a seam in paneling will occur to hide joints due to shrinkage.
      - Cut the first sheet to a length.
      - Plumb the edge and tack it temporarily into position (figure A).
      - Set the distance between the points of a scriber to the same as the amount the sheet overlaps the center of the stud. Scribe this amount on the edge of the sheet butting the corner (figure B).
      - Plane the edge to the line to complete the cut.
      - Replace the sheet with the cut edge fitting snugly in the corner.
   b. Wall Outlets
      - Plumb and mark both sides of the outlet.
      - Level and plumb marks from the wall and floor onto the sheet for the location of the opening (figure A).
      - When using a saber saw, cut from the back of the panel to avoid splintering the face (figure B).
   c. Fastening
      - Apply adhesive beads to the back of the panel.
      - Press the panel surface to the wall to make contact with the adhesive.
      - Slowly pull the bottom of the sheet a few inches away from the wall.
      - Press the sheet back into position after about two minutes.
      - Drive nails as needed.
   d. Ending the Application
      - Take measurements at the top, center, and bottom.
      - Rip the panel about 1/2 inch wider than the greatest distance.
      - Place the sheet plumb with the cut edge in the corner and the other edge overlapping the last sheet installed.
      - Set the scriber for the amount of overlap.
      - Scribe this amount on the edge in the corner (figure A).
      - Exterior corners may be finished by capping the joint (figure B).
8. Installing Solid Wood Paneling
   a. Starting the Application
      - Select a straight board to start with.
      - If tongue-and-grooved stock is used, position with the grooved edge in the corner.
      - Scribe.
      - Rip and plane to the scribed line.

- Replace and face nail along the cut edge.
- Blind nail the other edge through the tongue (figure A).
- Apply succeeding boards by blind nailing into the tongue only.

b. Applying the Last Board

- Cut and fit the next-to-last board.
- Remove it.
- Tack the last board in the place of the next-to-last board just previously removed.
- Cut a scrap block about 6 inches long and equal in width to the finished face of the next-to-last board.
- Use this block to scribe the last board.
- Cut and plane it to the scribed line.
- Fasten the next-to-last board in position.
- Fasten the last board in position.
- Face nail the edge nearest the corner.

Chapter 15 Interior Finish

Instructor's Note

In this column, you will find a variety of lecture tips to aid student comprehension, notes on when to use PowerPoint slides to teach concepts, safety cautions for your students, and notes on when your class should view a video segment.

Appendix A lists corresponding numbered PowerPoint Presentation titles (see page 197)

Appendix B lists corresponding numbered Video titles and sections as applicable (see page 201)

Introduction

1. Interior finish is a term used to describe the process of completing the interior of a building.
2. It collectively refers to installing the finished floor, finished ceiling, and moldings.
3. A finished ceiling may be created by installing a suspended ceiling.
4. Floors may be finished with wood, which is a long-time favorite because of its durability, beauty, and warmth.

OBJECTIVES

- Identify the components of a suspended ceiling system.
- Lay out and install suspended ceilings.
- Identify standard interior moldings and describe their use.
- Apply ceiling and wall molding.
- Apply interior door casings, baseboard, base cap, and base shoe.
- Install window trim, including stools, aprons, jamb extensions, and casings.
- Apply strip and plank flooring.
- Estimate quantities of the parts in a suspended ceiling system.
- Estimate the quantities of molding needed for windows, doors, ceilings, and base.
- Estimate wood flooring required for various installations.

Glossary of Interior Finish Terms

Stool – the bottom horizontal member of interior window trim that serves as the finished window sill

Apron – a piece of the window trim used under the stool

Compound miter – a bevel cut across the width and also through the thickness of a piece

Spline – a thin, flat strip of wood inserted into the grooved edges of adjoining pieces

A. Ceiling Finish
   1. Suspended ceilings provide space for recessed lighting, ducts, and pipes (figure 15-1).
   2. Easily installed in basements.
   3. Suspended Ceiling Components
      a. Wall angles

- L-shaped pieces that are fastened to the wall to support the other components of the ceiling system.
- Come in 10- and 12-foot lengths.

b. Main Runners
- Come in 12-foot lengths.
- End splices make it possible to join lengths of main runners together.
- Slots are punched in the side of the run at 6- or 12-inch intervals to receive cross tees.
- Punched holes are spaced at intervals for suspending main runners with hanger wire.

CAUTION Regarding cross tees: Care should be observed when working with the metal grid system around electrical wiring. The entire grid could become charged with one misplaced wire.

c. Cross Tees
- Come in 2- and 4-foot lengths.
- Come with connecting tabs on each end which lock into main runners and other cross tees.

d. Ceiling Panels
- Panel selection is based on considerations such as fire resistance, sound control, thermal insulation, light reflectance, moisture resistance, maintenance, appearance, and cost.
- Are available in 2 foot × 2 foot and 2 foot × 4 foot sizes with square or rabbeted edges (figure 15-3).

4. Suspended Ceiling Layout
   a. Sketch a grid plan by first drawing the overall size of the ceiling to a convenient scale.
   b. Locating Main Runners
      - Change the width of the room to inches and divide by 48.
      - Add 48 inches to any remainder.
      - Divide the sum by 2 to find the distance of the cross tee from the wall.
5. Constructing the Grid Ceiling System
   a. Install wall angles.
   b. Install suspended ceiling lags and hanger wires, suspending the main runners.
   c. Insert full-length cross tees.
   d. Cut and insert border cross tees.
   e. Suspended ceiling must be installed with at least 3 inches for clearance below the lowest air duct, pipe, or beam.
   f. For step-by-step instructions on constructing the grid ceiling system see the Procedures section.

B. Interior Molding

1. Manufactured in several sizes and patterns.
2. Standard patterns are usually made only from softwood.
3. Finger-jointed lengths are used when a paint finish is to be applied.
4. Standard Molding Patterns
   a. Standard moldings are designated as bed, crown, cove, full round, half round, quarter round, base, base shoe, base cap, casing, chair rail, back band, apron, stool, stop, and others (figure 15-8).
   b. Come in lengths of 8, 10, 12, 14, and 16 feet.
5. Molding Shape and Use – some moldings are classified by the way they are shaped. Others are designated by location.
   a. Quarter rounds may be used to trim inside corners (figure 15-10).

   b. Base shoe is normally used to conceal the joint between the bottom of the base and the finish floor (figure 15-11).
   c. Casings are used to trim around windows, doors, and other openings.
   d. Back bands are applied to the outside edges of casings for a more decorative appearance.
   e. Aprons, stools, and stops are parts of window trim.
   f. Outside corners are used to finish exterior corners of interior wall finish. Caps and chair rail trim the top edge of wainscoting.

6. Making Joints on Molding
   a. Joints on exterior corners are mitered.
   b. Joints on interior corners are usually coped.
   c. The most popular way to cut miters and other cuts is with a power miter box (figure 15-14).
   d. All moldings are cut with their face sides or finished edges up toward the operator.
   e. Bed, crown, and cove moldings require a compound miter, which may be cut two ways.
      - Compound miter saw.
      - Positioning the trim piece upside down in the miter box (figure 15-17).
   f. Making a Coped Joint
      - Make a miter cut on the end as if the joint was to be an inside miter (figure 15-18).
      - Use a coping saw cut along the outlined profile with a slight undercut (figure 15-19).
7. Installation Thoughts
   a. Interior trim should be sanded smooth after it has been cut.
   b. Exposed corners of trim should be rounded over slightly.
   c. If the trim is to be stained, make sure every trace of glue is removed.
   d. Be careful not to make hammer marks in the finish.
   e. Make sure any pencil lines left along the edge of a cut are removed.
   f. Make sure all joints are tight-fitting.
8. Applying Wall Molding
   a. For large size ceiling molding, such as beds, crowns, and coves, a chalk line ensures straight application of the molding and easier joining of the pieces.
   b. For step-by-step instructions on applying wall molding, see the Procedures section.

VIDEO #7
Section 5

9. Door Casings
   a. Design of Door Casings
      - When molded casings are used, the joint at the head must be mitered unless butted against plinth blocks (figure 15-21).
      - When using S4S lumber, the joint may be mitered or butted.
      - Molded casings may be enhanced with the application of back bands (figure 15-22).
      - Backed out surfaces allow the casing to come up tight on both wall and jamb (figure 15-23).
   b. Applying Door Casings
      - Door casings are set back from the inside face of the door face a distance of about 5/16 inch.
      - This setback is called a reveal (figure 15-24).

- Cut side casings in pairs with right- and left-hand miters for use on both sides of the opening.
- Use 4d finish nails on inside of casing.
- Use 6d or 8d finish nails on outside edge.

VIDEO #7
Section 9

10. Base Trim
    a. Molded or S4S stock may be used for base.
    b. Base cap may be applied to the top edge.
    c. For step-by-step instruction on applying base moldings, see the Procedures section.

VIDEO #7
Section 6

11. Window Trim
    a. Consists of the stool, apron, jamb extensions, and casings (figure 15-25).
    b. Stool
       - May be rabbeted at an angle to fit on the window sill.
       - Each end projects beyond the casings by an amount equal to the casing thickness.
    c. Extension jambs
       - Extend the jamb inside edges flush to the inside wall surface.
    d. Window casings usually are installed with a reveal similar to that of door casings.
    e. For step-by-step instructions on installing window trim, see the Procedures section.

C. Finish Floors

1. Species
   a. Hardwood finish flooring is made from white or red oak, beech, birch, hard maple, and pecan.
   b. Some softwoods such as Douglas fir, hemlock, and southern yellow pine are used.
2. Kinds of Solid Hardwood Flooring
   a. Strip Flooring
      - Strip flooring is probably the most widely-used type.
      - Most strips are tongue-and-grooved on edges.
      - Unfinished.
      - Prefinished.
      - The most popular size of hardwood strip flooring is 3/4 inch thick with a face width of 2 1/4 inches (figure 15-26).
   b. Plank Flooring
      - Solid wood plank flooring is similar to strip flooring.
      - Comes in various width combinations ranging from 3 to 8 inches.
3. Grades of Hardwood Flooring
   a. Uniform grading rules have been established for strip and plank solid hardwood flooring by the National Oak Flooring Manufacturers Association.
   b. Oak flooring is available quarter-sawed and plain-sawed.
      - Grades in declining order are clear, select, no. 1 common, no. 2 common, and 1 1/4 foot shorts.
      - Quarter-sawed flooring is available in clear and select grades only.
   c. Birch, beech, and hard maple flooring are graded in declining order as first grade, second grade, third grade, and special grade.
   d. In addition to appearance, grades are based on length.

4. Laying Wood Floor
   a. General
      - In new construction, the base or door casings are not usually applied to allow for easier application of the finish floor.
      - In remodeling, the base and base shoe must be removed and the casing bottoms cut off.
      - The flooring may be laid in either direction on a plywood or diagonal board subfloor. But it must be laid perpendicular to subfloor boards that run at right angles to the joists (figure 15-28).

   b. Handling and Storage of Wood Finish Floor
      - Flooring should not be delivered to the job site until the building has been closed in.
      - Cement work, plastering, and other materials must be thoroughly dry. In warm seasons, the building should be well ventilated.
      - During cold months, the building should be heated, not exceeding 72 degrees Fahrenheit.
      - Let the flooring become acclimated to the atmosphere of the building for four or five days or more before installation.

   c. Concrete Slab Preparation
      - Wood floors should not be installed on below-grade slabs.
      - New slabs should be at least ninety days old.
      - Flooring should not be installed when tests indicate excessive moisture in the slab.
      - Grind off any high spots.
      - Fill low spots with leveling compound.
      - A moisture barrier must be installed over all concrete slabs.
        1) Spread a skim coat of mastic with a straight trowel over the entire slab.
        2) Then cover the slab with polyethylene film.
      - A plywood subfloor may be installed over the moisture barrier on which to fasten the finish floor.
        1) Leave a 3/4-inch space at walls and 1/4- to 1/2-inch space between panel edges and ends (figure 15-29).
      - Finish flooring may also be fastened to sleepers installed on the slab.
      - The edges are lapped over the rows (figure 15-30).

   d. Blind Nailing
      - At about a 45-degree angle, start the nail in the corner at the top edge of the tongue.
      - Recommendations for fastening are shown in figure 15-31.
      - For the first two or three courses of flooring, a hammer must be used to drive the fasteners.
      - A nail is laid on edge with its head on the nail to be set. With one sharp blow, the nail is set (figure 15-33).
      - A nail set should not be used to set hardened flooring nails.

   e. Using the Power Nailer
      - Although it is called a power nailer, a heavy hammer is swung by the operator against a plunger to drive the fastener (figure 15-34).
      - The hammer is double-headed. One head is rubber and the other is steel.

CAUTION Eye protection should be worn whenever driving nails, particularly hardened nails. A small piece of steel may break off the hammer or the nail and fly out in any direction. This could cause eye injury.

f. Ending the Flooring
   - The last three or four courses from the opposite wall must be nailed by hand.
   - Use a bar to pry the pieces tight at each nail location (figure 15-36).

g. Use flooring, with mitered joints at the corners, as framework around the obstructions (figure 15-38).

h. Change directions by joining groove edge to groove edge and inserting a spline, ordinarily supplied with the flooring (figure 15-39).

i. Installing Plank Flooring
   - Plank flooring is installed like strip flooring.
   - Alternate the courses by widths.

D. Installing Closet Trim

1. A simple clothes closet is normally furnished with a *shelf* and a *rod* for hanging clothes.
2. A piece of 1 × 4 or 1 × 6 stock is installed around the walls of the closet to support the shelf and the rod.
3. The height from the floor to the top edge of the cleat is typically 66 inches (figure 15-40).

E. Linen Closets

1. Linen closets usually consist of a series of shelves spaced 12 to 16 inches apart.

F. Mantels

1. Mantels are used to decorate fireplaces and to cover the joint between the fireplace and the wall.
2. Most mantels are made offsite or come preassembled from the factory. They are available in a number of sizes and styles (figure 15-41).
3. Study the manufacturer's directions carefully. Place the mantel against the wall. Center it on the fireplace. Scribe it to the floor or wall as necessary. Carefully fasten the mantel in place and set all fasteners.

G. Estimating

1. Suspended Ceiling

   a. Wall Angle – divide the perimeter of the room by the length of a wall angle.

   b. Main Runners
      - Divide the length of the room by 12 feet.
      - Round up.
      - Multiply this quantity by the number of main runner rows needed.

   c. Cross Tees
      - Divide the room length by 2 feet.
      - Round this number down for 4-foot tees and round it up for 2-foot tees.
      - Multiply these numbers by the number of main runner rows.

   d. Suspended Ceiling Lags – number of lags is equal to one-half the number of 4-foot cross tees.

   e. Hanger Wire
      - Add 1 foot to the distance between the suspended ceiling and the ceiling supports.
      - Multiply this length by the number of lags.

   f. 2 × 4 Ceiling Panels
      - Divide the room length by 2 and round up.
      - Divide the room width by 4 and round up.

- Multiply these numbers together.
- Deduct one for each panel-size light fixture.

2. Interior Molding
   a. Window Trim
      - Each window is figured separately and then added together.

   b. Door Casing – Head casing length is a little more than the door width plus two casing widths. Side casing lengths are a little longer than the door height plus one casing width.
   c. Base Molding – From the room perimeter, deduct the widths of each door opening.
   d. Ceiling Molding – Divide the room perimeter by the length of the molding being purchased.
3. Hardwood Flooring
   a. Determine the area to be covered.

H. Procedures

1. Constructing the Grid Ceiling System
   a. Locate the height of the ceiling.
   b. Snap chalk lines on all walls around the room to the height of the top edge of the wall angle.
   c. If a laser is used the chalk line is not needed since the ceiling is built to the light beam (figure A).
   d. Fasten wall angles around the room.
   e. Determine the position of the first main runner. Stretch a line at this location across the room from the top edges of the wall angle.
   f. Install the cross tee line.
   g. Install hanger lags not more than 4 feet apart and directly over the stretched line.
   h. Cut a number of hanger wires using wire cutters.
   i. Attach the hanger wires to the hanger lags.
   j. Then make a 90-degree bend in the wire where it crosses the stretched line (figure B).
   k. At each main runner location, measure from the wall to the cross tee line.
   l. Hang the main runners by resting the cut end on the wall angle and inserting suspension wires in the appropriate holes in the top of the main runner (figure C).
   m. Cross tees are installed by inserting the tabs on the ends into the slots in the main runners (figure D).
   n. Lay in a few full-size ceiling panels to stabilize the grid while installing the border cross tees (figure E).
   o. Cut and install cross tees along the border (figure F).
   p. For 2 × 2 panels, install 2-foot cross tees at the midpoints of the 4-foot cross tees.
   q. Ceiling panels are placed in position by tilting them slightly, lifting them above the grid, and letting them fall into place (figure G).
2. Applying Wall Molding
   a. Snap a line for wall trim (figure A).
   b. Apply the molding to the first wall with square ends in both corners.

c. Cope the starting end of the first piece on each succeeding wall against the face of the last piece installed on the previous wall.

d. The end of the last piece installed must be coped to fit against the face of the first piece.

VIDEO #7
Section 5

SLIDE 32

Mark the bottom end by turning it upside down with the point of the miter touching the floor. Mark the side casing in line with the top edge of the head casing. Make a square cut on the casing at that mark (figure 15-54).

3. Applying Door Casings
   a. Set the blade of the combination square so that it extends 5/16 inch beyond the body of the square.
   b. Gauge lines at intervals along the side and head jamb edges.
   c. Let the lines intersect where side and head jambs meet (figure A).
   d. Cut a miter on the ends of the four side casings in pairs of right and left members.
   e. Miter head casing.
   f. Fasten the head casing in position.
   g. The mitered ends should be in line with the gauged lines on the side jambs.
   h. Cut the previously mitered side casing to length.
   i. When fitted, a little glue may be applied to the joint.

SLIDE 33

VIDEO #7
Section 9

4. Applying Base Moldings
   a. Cut the first piece with squared ends if it fits between two walls. Miter the butt joint if desired (figure A).
   b. Fasten in place.
   c. Make regular miter joints on outside corners.
   d. If a base cap is applied, it is done so in the same manner as most wall or ceiling molding.
   e. It should be nailed into the floor and not into the baseboard (figure B).
   f. Generally, no base shoe is required if carpeting is to be used as a finish floor (figure C).

VIDEO #7
Section 9

SLIDE 34

Raise the lower sash slightly. Place a short, thin strip of wood under it, on each side, which projects inward to support the stool while it is being laid out (figure 15-59). Place the stool on the strips. Raise or lower the sash slightly so the top of the stool is level.

5. Installing Window Trim
   a. Applying the Stool
      - Hold a piece of side casing in position at the bottom of the window and draw a light line on the wall along the outside edge of the casing stock.
      - Mark a distance outward from these lines equal to the thickness of the window casing.
      - Cut a piece of stool stock to length equal to the distance between the outermost marks.
      - Position the stool with its outside edge against the wall (figure A).
      - Lightly square lines across the face of the stool, even with the inside face of each side jamb of the window frame (figure B).
      - Set the pencil dividers or scribers to mark the cutout so that, on both sides, an amount equal to twice the casing thickness will be left on the stool.
      - Cut to the lines, using a handsaw.
      - Apply a small amount of caulking compound to the bottom of the stool. Fasten the stool in position.

VIDEO #7
Section 8

   b. Applying the Apron
      - Cut a length of apron stock equal to the distance between the outer edges of the window casings.
      - Each end of the apron is then returned upon itself (figure A).
   c. Installing Jamb Extensions
      - Measure the distance from the jamb to the finished wall.

- Rip the jamb extensions to this width with a slight back-bevel on the side towards the jack stud (figure A).

VIDEO #7
Section 7

d. Applying the Casings
   - Cut side casings with left- and right-hand miters.
   - Install the header casing first and then the side casings in a similar manner to door casing.

SLIDE 35

6. Installing Wood Flooring
   a. Preparing for Installation
      - Check the subfloor.
      - Cover the subfloor with building paper.
      - Snap chalk lines on the paper showing the centerline of floor joists.
   b. Starting Strip
      - Place a strip of flooring on each end of the room, 3/4 inch from the starter wall with the groove side toward the wall.
      - Mark along the edge of the flooring tongue.
      - Snap a chalk line between the two points.
      - Hold the strip with its tongue edge to the chalk line.
      - Make sure end joints between strips are driven up tight (figure A).
      - Cut the last strip to fit loosely against the wall.
   c. Racking the Floor
      - Lay out seven or eight loose rows of flooring, end to end (figure B).
      - End joints should be at least 6 inches apart.
      - Find or cut pieces to fit within 1/2 inch of the end wall.
      - Rack seven or eight courses ahead as work progresses.
7. Installing Closet Shelves
   a. Measure up from the floor the height of the cleat, usually 66 inches. Draw a level line on the back wall and two end walls of the closet.
   b. Drill the closet pole sockets in the end cleats. The center of the socket should be at least 12 inches from the back wall and centered on the width of the cleat.
   c. Cut the shelf about 1/2 inch longer than the distance between end walls (figure A).
   d. Place the shelf in position by laying one end on the cleat and tilting the other end up and resting against the wall. Scribe about 1/4 inch off the end resting on the cleat.
   e. Measure the distance between corners along the back wall.
   f. Place the shelf in position, tilted in the opposite direction on the cleat.
   g. Set the dividers to scribe the distance from the wall to the mark on the back edge of shelf.
   h. Replace the shelf into position, resting it on the cleats.
   i. Scribe the back edge to the wall to take off as little as possible.
   j. Measure the distance between pole sockets, adjusting the pole length to include the depth of the sockets. Cut the pole to length.

Chapter 16 Stair Framing and Finish

Instructor's Note

In this column, you will find a variety of lecture tips to aid student comprehension, notes on when to use PowerPoint slides to teach concepts, safety cautions for your students, and notes on when your class should view a video segment.

Appendix A lists corresponding numbered PowerPoint Presentation titles (see page 197)

Appendix B lists corresponding numbered Video titles and sections as applicable (see page 201)

Introduction

1. The staircase is usually one of the most outstanding features of a building's interior.
2. A high degree of skill is necessary for the design and construction of staircases.
3. It is important that staircases be comfortable and safe to use.
4. Many kinds of stair finish parts are manufactured in a wide variety of wood species, such as oak, beech, cherry, poplar, pine, and hemlock.
5. It is important to identify each of the staircase parts, know their locations, and understand their functions.
6. It is helpful to know that all stairs are laid out using the same theory. This theory is also similar to that used for rafter layout.
7. The variations in stair construction depend largely on the stair function, location, and component material used.

OBJECTIVES

- Name various stair finish parts and describe their location and function.
- Describe several stairway designs.
- Define terms used in stair framing.
- Determine the unit rise and unit run of a stairway given the total rise.
- Determine the length of a stairwell.
- Lay out a stair carriage and frame a straight stairway.
- Lay out and frame a stairway with a landing.
- Lay out, dado, and assemble a housed-stringer staircase.
- Apply finish to the stair body of open and closed staircases.
- Install a post-to-post balustrade from floor to balcony on the open end of a staircase.

Glossary of Stair Framing Terms

Balustrade – the entire stair rail assembly, including handrail, balusters, and posts

Baluster – vertical members of a stair rail, usually decorative and spaced closely together

Handrail – a railing on a stairway intended to be grasped by the hand to serve as a support and guard

Newel post – an upright post supporting the handrail in a flight of stairs

Rake – the sloping portion of trim, such as on gable ends of a building or stair

Shank hole – a hole drilled for the thicker portion of a wood screw

Stairwell – an opening in the floor for climbing or descending stairs, or the space of a structure where the stairs are located

A. Stairs

1. Stairway Design
   a. General
      - The stairwell is the opening in the floor through which a person must pass when climbing and descending the stairs (figure 16-1).
      - Stairs are further defined as finish or service stairs.
      - Stairways in residential construction should be at least 36 inches wide (figure 16-2).

   b. Types of Stairways
      - A straight stairway is continuous from one floor to another.
      - Platform stairs have intermediate landings between floors.
      - An L-type platform stairway changes direction 90 degrees.
      - A U-type platform stairway changes direction 180 degrees.
      - A winding staircase gradually changes direction as it ascends from one floor to another.
        1) Their use is not recommended (figure 16-3).
      - Stairways constructed between walls are called closed stairways.
      - Stairways that have one or both sides open to a room are called open stairways.

VIDEO #8 Section 2

2. Stair Framing Terms
   a. Total Rise – the vertical distance between finish floors.
   b. Total Run – the total horizontal distance that the stairway covers.
   c. Unit Rise – the vertical distance from one step to another.
   d. Riser – the finish material that covers this distance.
   e. Tread – the horizontal member on which the feet are placed when climbing or descending the stairs.
   f. Unit Run – the horizontal distance between the faces of the risers along the tread.
   g. Nosing – that part of the tread that extends beyond the face of the riser.
   h. Stair Carriage – the main support under the risers and treads.
   i. Stair Stringer – serves as a finish or trim piece to the stairs that covers the carriage. It may also serve as the main support for the risers and treads when a carriage is not used.
   j. Stairwell – an opening in the floor for the stairway to pass through.
   k. Headroom – the smallest vertical distance between the stairs and the upper construction over the foot of the stairs.
   l. Closed Staircase – one where the treads and risers end against a vertical surface.
   m. Open Staircase – when the ends of the tread and risers are visible.

SLIDE 36

VIDEO #8 Section 3

3. Determining the Unit Rise and Unit Run
   a. Staircases must be constructed at an appropriate angle for maximum ease in climbing and for safe descent.

Given floor-to-floor measurements, have students calculate appropriate height and number of risers.

b. The height of a riser shall not exceed 7 3/4 inches.

c. The width of a tread should not be less than 10 inches.

d. Determine unit rise

- Measure the total rise of the stairway.
- Find the number of risers that will fit in the opening.
- Dividing the total rise again by the number of rises gives the riser height.

Given maximum available total run, have students calculate acceptable tread amounts and dimensions.

e. Tread Run

- The sum of one unit rise and one unit run should equal between 17 and 18.
- Another formula to find the unit run states that the sum of two risers and one tread shall not be less than 24 inches nor more than 25 inches.

f. Decreasing the unit rise increases the run of the stairs.

g. Increasing the riser height decreases the run.

h. A riser height of 7 1/2 inches and a unit run of 10 inches makes a safe, comfortable stairway.

4. Determining the Size of the Stairwell

Have students calculate stairwell lengths for previously calculated staircases.

a. Stairwell Width

- The stairwell must be made wider than the staircase to allow for wall and stair finish (figure 16-8).
- Extra width will be required for a handrail and other finish parts of an open staircase that makes a U-turn on the landing above.

SLIDE 37

b. Length at the Stairwell

- The length of the stairwell depends on the slope of the stairway (figure 16-9).
- Most building codes require a minimum of 6′ 8″ for headroom.
- To find the minimum length of the stairwell
  1) Add the thickness of the floor assembly above (sum of subfloor, floor joists, and ceiling finish) to the desired headroom.
  2) Divide this number by the unit rise and round up to the next largest whole number.
  3) Then multiply this number by the unit run to find the minimum length of the stairwell.
- Additional headroom can be obtained by framing the header above the low end of the staircase at the same angle as the stairs (figure 16-11).

5. Methods of Stair Construction

a. Open Carriage Staircases

- The open carriage staircase is laid out for cutouts.
- Risers and treads are fastened to the cutouts (figure 16-12).

b. The Housed Stringer Staircase

- Dadoes are routed into the sides of the finish stringer (figure 16-13).
- A router and stair jig are used to dado the stringers.
- Stair carriages are not required when the housed finish stringer method of construction is used.

c. Determining the Rough Length of a Staircase

- May be determined by using the Pythagorean Theorem.
- Can also be found by scaling across a framing square.
  1) Use the edge of the square that is graduated in 12ths of an inch.
  2) Mark the total rise on the tongue.

3) Then mark the total run on the blade.
4) Scale off in between the marks (figure 16-14).

d. Laying Out and Framing a Stairway with a Landing
- A landing is designed for changing the direction of the stairs and as a resting place for long stair runs (figure 16-15).
- Many codes state that no flight of stairs shall have a vertical rise of more than 12 feet.
- Codes require the minimum dimension to be the width of the stairway. U-type stairs usually have the landing about the midway on the flight. Platform stairs are built by first erecting the platform (figure 16-16).

SLIDE 38

B. Stair Carriage

1. General

VIDEO #8
Sections 4–6

a. When laying out stair carriages, make sure that every riser height will be the same.
b. Make sure all tread widths will be equal when the staircase is finished.

2. Layout

Cover 2″ × 12″ stock with kraft paper to practice stair layout.

a. Place the stair carriage stock on a pair of sawhorses.
b. Sight the stock for a crowned edge.
c. Set stair gauges on the framing square with the unit rise on the tongue and the unit run on the blade.
d. Lines marked out along the tongue will be plumb lines and those along the blade will be level lines.
e. Step-off the necessary number of times, marking along the outside of the tongue and blade (figure 16-17).
f. Lay out a level line at the top of the carriage for the tread.
g. Lay out a level line on bottom where the carriage sits on the floor.
h. A certain amount may have to be cut off the bottom end of the stair carriage.
   - If the carriage rests on the finish floor, the first riser is reduced by the thickness of the tread stock.
   - If the bottom of the carriage rests on the subfloor, and the finish floor and tread stock are the same thickness, the height of the first riser is the same as all the rest.
   - If the bottom of the carriage rests on the subfloor and the tread stock is thicker than the finish floor, then the first riser is reduced by the difference of tread and finished floor thicknesses (figure 16-18).
i. No amount is cut from the level line at the top of the carriage.

VIDEO #8
Section 7

j. Allowances must be made so that the top finished tread width will be the same as all others in the staircase (figure 16-20).

3. Cutting the Stair Carriages

Use 7/16″ OSB ripped at 9 1/4″ for an economical alternative to dimensional lumber when cutting stair carriages.

a. Follow the layout lines carefully.
b. Using the first carriage as a pattern, lay out and cut as many other carriages as needed.

4. Installing the Stair Carriage

VIDEO #8
Section 8

a. Fasten the first carriage in position on one side of the stairway.
b. Place nails near the bottom edge of the carriage to prevent splitting the triangular sections.
c. Fasten a second carriage on the other wall in the same manner as the first.

   d. If the stairway is to be open on one side, position the stair carriage on the open side of staircase, making sure its outside face will be plumb with the centerline of the handrail when it is installed (figure 16-21).
   e. Fasten intermediate carriages (figures 16-22 and 16-23).

VIDEO #8
Sections 9, 10

5. Finishing a Closed Stair Carriage
   a. Risers
      - Risers are installed first (figure 16-24).
      - Rip the riser stock to the proper width.
      - Cut the risers to length, with square ends, about 1/4 inch less than the distance between walls.
      - Fasten the risers in position with three 1 1/2-inch finish nails into each stair carriage.
   b. Closed Finish Stringer
      - The closed finish stringer is cut around the previously installed risers (figure 16-25).
      - Tack a length of stringer stock to the wall.
         1) Its bottom edge should rest on the top edges of the previously installed risers and its bottom ends should rest on the floor.
         2) The top end should extend about 6 inches beyond the landing.
      - Lay out plumb lines, from the face of each riser, across the face of the finish stringer.
      - Lay out level lines on the stringer from each tread cut of the stair carriage and also from the floor of the landing above.
      - Remove the stringer from the wall.
      - Cut to the layout lines.
      - Sand the board, and place it back in position.
      - Fasten the stringer securely to the wall with finishing nails.
      - Shim the risers out tight to the stringer.
   c. Treads

Rubbing the ends with wax will help when placing the tread into position.

VIDEO #8
Sections 9, 10

      - Treads are cut on both ends to fit snugly between the finish stringers.
      - The nosed edge of the tread projects beyond the face of the riser by 1 1/8 inches (figure 16-26).
      - Cut to length carefully.
      - Using a scrap block on the nosed edge, tap the tread into position until the back edge is firmly against the riser.
      - If it is not possible to work from the underside, the treads must be face nailed.
      - At the top of the stairs, install a landing tread (figure 16-27).
   d. Tread Molding
      - The tread molding is installed under the overhang of the tread and against the riser.
      - Cut the molding to the same length as the treads, using a miter box.
      - Fasten the molding in place with 4d finish nails.
6. Finishing an Open Stair Carriage
   a. Finish Stringers
      - The open finish stringer must be installed before the risers and the closed finish stringer.
      - Cut a length of finish stringer stock.
      - Fit it to the floor and against the landing.

- Its top edge should be flush with the top edge of the stair carriage.
- Lay out level lines on the face of the stringer in line with the tread cut on the stair carriage.
- Plumb lines must be laid out on the face of the finish stringer for making miter joints with risers.
- Use a preacher to lay out the plumb lines on the open finish stringer.
- Lay out the plumb cut on the stringer by marking along the side of the preacher that faces the bottom of the staircase (figure 16-28).

b. Risers
- Cut risers to length by making a square cut on the end that goes against the wall.
- Make miters on the other end to fit the mitered plumb cuts of the open finish stringer (figure 16-29).
- Fasten them in position to each stair carriage.

c. Treads
- Rip the treads to width. Cut one end to fit against the closed finish stringer.
- Make a cut on the other end to receive the return nosing. This is a combination square and miter cut.
  1) The square cut is made flush with the outside face of the open finish stringer.
  2) The miter distance is equal to the width of the return nosing beyond the square cut.

d. Return Nosings
- The return nosings are applied to the open ends of the treads.
- Miter one end of the return nosing to fit against the miter on the tread.
- Cut the back end square.
- Return the end on itself (figure 16-30).

e. Tread Molding
- It is mitered on the open end and returned back onto the open stringer.
- The back end of the return molding is cut and returned upon itself at a point so the end assembly shows the same as at the edge (figure 16-31).

C. Housed Stringer

1. General
   a. Housed stringers have risers and treads set into dados made in the stringer.
   b. Then the treads and risers are wedged tightly against the shoulders of the dadoes (figure 16-32).
   c. These dados are made by using a stair template to guide a router (figure 16-33).
2. Housing the Stringer
   a. On the face side of the stringer stock, lightly draw a line parallel to and about 2 inches down from the top edge.
   b. Lay out the risers and treads for two or three steps of the staircase (figure 16-34).
   c. Lightly square lines to the top edge of the stringer at the intersection of the face sides of tread and riser (figure 16-35).
   d. Set the template over the first step and adjust it to match the layout.
   e. Rout the stringer about 1/4 inch deep.
   f. Cut and fit the bottom end of the stringer to the floor and the top end to the landing.

3. Laying Out an Open Stringer
   a. Mark lightly the riser and tread layouts on the stringer. The unit rise and unit run intersect on the top edge of the stringer.
   b. Lay out the miter cut for the risers.
   c. Measure perpendicular to the riser layout line a distance equal to the thickness of the riser stock. Draw another plumb line at this point.
   d. Square both plumb lines across the top edge of the stringer stock.
   e. Draw a diagonal line on the top edge to mark the miter angle (figure 16-36).
   f. Mark the tread cut line in the stringer.
   g. Measure down from the tread layout line a distance equal to the thickness of the tread stock.
   h. Draw a level line at this point for the tread cut.
   i. The tread cut is square through the thickness of the stringer.
   j. Fit the bottom end of the stringer to the floor.
   k. Fit the top end against the landing.
   l. Make the mitered plumb cuts for the rises and the square level cuts for the treads.
4. Installing Risers and Treads
   a. Rip and cut the treads to width and length.
   b. On the open side of a staircase, a miter joint is made so no end grain is exposed (figure 16-37).
   c. The risers are installed with wedges, glue, and screws between housed stringers.
   d. The treads are then installed with wedges, glue, and screws on the closed side.
5. Applying Return Nosings and Tread Molding
   a. If the staircase is open, a return nosing is mitered to the end of the tread.
   b. The back end of the return nosing projects past the riser the same amount as the tread overhangs the riser.
   c. The end is returned upon itself.
   d. The tread molding is then applied under the overhang of the tread.

D. Balustrade

1. Balustrade Members
   a. Newel Posts
      - Starting newels are used at the bottom of a staircase.
      - Intermediate landing newels (figures 16-38, 39, and 40).
   b. Handrails
      - Handrail heights are typically 30 to 38 inches vertically above the nosing edge of the tread.
      - There should be continuous 1-1/2 inch finger clearance between the rail and the wall.
      - Some handrails are plowed with a wide groove on the bottom side to hold square-top balusters in place (figures 16-42 and 43).
      - On closed staircases, a balustrade may be installed on top of a kneewall or buttress.
   c. Balusters
      - They are spaced close together and support the handrail.
      - On a kneewall, they run from the handrail to the shoe rail.
      - On an open staircase, they run from the handrail to the treads (figure 44).

2. Laying Out Balustrade
   a. Its centerline is first laid out (figure 16-45).
   b. Lay Out Baluster Centers
      - Most codes require that balusters be spaced so that no object 4 inches in diameter or greater can pass through.
      - If two balusters are used on each tread, the spacing is half the run.
      - If codes require three balusters per tread, the spacing is one-third the run (figure 16-46).
3. Installing a Balustrade
   a. Laying Out the Handrail
      - Clamp the handrail to the tread nosings.
      - Clamp at the tread nosing to avoid bowing the handrail.
      - Use a framing square to mark the handrail where it will fit between starting and the balcony or upper newel post (figure 16-47).
      - Use a framing square at the landing nosing to measure the vertical thickness of the rake handrail.
      - At the bottom, measure the height from the first tread to the top of the handrail where it butts the newel post.
      - Record and save the measurements for later use (figure 16-48).
   b. Determining the Height of the Starting Newel
      - The height of the stair handrail is taken from the top of the tread along a plumb line flush with the face of the riser (figure 16-49).
      - Stairways wider than 44 inches require a handrail on both sides.
      - Handrail heights are typically 30 to 38 inches vertically above the nosing edge of the tread.
      - If a turned starting newel post is used, add the difference between the two previously recorded measurements above to the required rake handrail height.
      - Then add 1 inch for a block reveal (figure 16-50).
   c. Installing the Starting Newel
      - Trim the starting newel to the desired length.
      - The starting newel is notched over the outside corner of the first step.
        1) One-half of its bottom thickness is left on from the front face of the post to the face of the riser.
        2) In the other direction, it is notched so its centerline will be aligned with the handrail centerline (figure 16-51).
      - Plumb it in both directions.
      - Secure in place with counterbored lag screws.
   d. Installing the Balcony Newel
      - Codes require that balcony rails for homes be no less than 36 inches.
      - Trim the balcony newel to the calculated height.
      - Plumb it in both directions.
      - Fasten it in place with counterbored lag bolts (figure 16-52).
   e. Boring Holes for Balusters
      - Bore holes in the treads at the center of each baluster (figure 16-53).
      - Cut the handrail to fit.
      - Place the handrail on the tread nosings between posts.

- Transfer the baluster centerlines from the treads to the handrail (figures 16-54 and 55).

f. Installing Handrail and Balusters

- Prepare the posts for fastening the handrail by counterboring and drilling shank holes for lag bolts through the posts.
- Temporarily fasten the handrail to the posts (figure 16-56).
- Cut the balusters to length.
- If square-top balusters are used, they are trimmed to length at the handrail angle and inserted into a plowed handrail (figure 16-57).
- Fillets are installed in the plow of the handrail, between the balusters.

4. Installing the Balcony Balustrade

a. Handrail

- Cut a half newel to the same height as the balcony newel (figure 16-58).
- Cut the handrail to length.
- Fasten the half newel to one end of it.
- Replace the half newel to the wall and temporarily fasten the other end of the handrail to the landing newel.
- A rosette may be used to end the handrail against a wall (figure 16-59).

b. Balcony Balusters

- The balcony balusters are spaced by adding the thickness of one baluster to the distance between the balcony newel and the half newel or wall.
- The overall distance is then divided into spaces that equal, as close as possible, the spacing of the rake balusters (figure 16-60).
- The balcony balusters are then installed in a manner similar to the rake balusters.

Chapter 17 Cabinets and Countertops

Instructor's Note

In this column, you will find a variety of lecture tips to aid student comprehension, notes on when to use PowerPoint slides to teach concepts, safety cautions for your students, and notes on when your class should view a video segment.

Appendix A lists corresponding numbered PowerPoint Presentation titles (see page 197)

Appendix B lists corresponding numbered Video titles and sections as applicable (see page 201)

Introduction

1. Cabinets and countertops complete the interior finish with useable workspace and storage.
2. They are often purchased in preassembled units and installed by carpenters.
3. Manufactured kitchen and bath cabinets come in a wide variety of styles, materials, and finishes (figure 17-1).
4. Cabinet making and assembly requires skill and patience to produce a finished product that is pleasing to look at as well as functional.
5. Workmanship is the grade stamp carpenters leave behind on their work.

OBJECTIVES

- State the sizes and describe the construction of typical base and wall kitchen cabinet units.
- Install manufactured kitchen cabinets.
- Construct, laminate, and install a countertop.
- Identify cabinet doors and drawers according to the type of construction and method of installation.
- Identify overlay, lipped, and flush cabinet doors and proper drawer construction.
- Apply cabinet hinges, pulls, and door catches.

Glossary of Cabinet and Countertop Terms

Face frame – a framework of narrow pieces on the face of a cabinet containing door and drawer openings

Gain – a cutout made in a piece to receive another piece, such as a cutout for a butt hinge

J-roller – a 3-inch wide rubber roller used to apply pressure over the surface of contact cement-bonded plastic laminates

Pilot – a guide on the end of edge-forming router bits used to control the amount of cut

Pivot – a point of rotation

Postforming – method used to bend plastic laminate to small radii

A. **Components of Manufactured Cabinets**
   1. Kinds and Sizes

a. General
   - One method of cabinet construction utilizes a face frame.
   - Another method, called European or faceless, eliminates the face frame (figure 17-2).
   - The surface of the countertop is usually about 36 inches from the floor.
   - Wall units are installed about 18 inches above the countertop.
   - The usual overall height of a kitchen cabinet installation is 7′-0″ (figure 17-3).

b. Wall Cabinets
   - Standard wall cabinets are 12 inches deep.
   - The standard height is 30 inches, but are also available in heights of 42, 24, 18, 15, and 12 inches.
   - Typical wall cabinet widths range from 9 to 48 inches in 3-inch increments.
   - Wall corner cabinets make access into corners easier.
   - Double-faced cabinets have doors on both sides for use above island and peninsular bases.

c. Base Cabinets
   - Most base cabinets are manufactured 34 1/2 inches high and 24 inches deep.
   - Single-door cabinets are manufactured in widths from 9 to 24 inches.
   - Double-door cabinets come in widths from 27 to 48 inches.
   - Some base units have no drawers; others contain all drawers.
   - Corner units, with round revolving shelves, make corner storage easily accessible (figure 17-5).

d. Tall Cabinets
   - Manufactured 24 inches deep.
   - Utility cabinets are 12 inches deep.
   - They are made 66 inches high and in widths of 27, 30, and 33 inches for use as oven cabinets.
   - Double-door pantry cabinets are made 36 inches wide (figure 17-6).

e. Vanity Cabinets
   - Most vanity cabinets are made 31 1/2 inches high and 21 inches deep.
   - Usual widths range from 24 to 36 inches in increments of 3 inches, then 42, 48, and 60 inches.

f. Accessories
   - Filler pieces fit small gaps between wall and base units when no combination of sizes can fill the existing space.
   - Other accessories include
     1. End panels.
     2. Face panels for dishwashers and refrigerators.
     3. Open shelves for cabinet ends.
     4. Spice racks.

2. Lay Out Kitchen Cabinets

   a. The blueprints for a building contain plans, elevations, and details that show the cabinet layout (figure 17-8).

   b. The wall cabinets usually match up with the base cabinets.

   c. Countertops are manufactured in various standard lengths.

   d. Most common are postformed countertops.

SLIDE 39

B. Installing Manufactured Cabinets

1. General
   a. Cabinets must be installed level and plumb even though floor and walls may not be.
   b. Level lines are first drawn on the wall for base and wall cabinets.
   c. If shimming base cabinets, lay out the level lines on the wall from the highest point on the floor where cabinets are to be installed. If fitting cabinets to the floor, measure up from the low point.
   d. The space between the top of the wall unit and the ceiling may be left open or it may be closed (figure 17-11).

C. Making a Countertop

1. Plastic Laminates
   a. Plastic laminates are widely used for surfacing kitchen cabinets and countertops.
   b. Laminates are manufactured in many colors and designs.
   c. Laminates come in widths of 24, 36, 48, and 60 inches and in lengths of 5, 6, 8, 10, and 12 feet.
   d. Sheets are usually manufactured 1 inch wider and longer than the nominal size.
   e. Contact bond adhesive is used.
2. Fitting the Countertop
   a. Place countertop panel material flat on the base cabinets, with its edge against the wall.
   b. Scribe the back edge of the countertop to the wall.
   c. Cut the countertop to the scribed line.
   d. Front edge should be flush with the face of the face frame (figure 17-12).
3. Applying the Backsplash
   a. Rip a 4-inch wide length of 3/4 inch stock the same length as the countertop.
   b. Fasten the backsplash on top of and flush with the back edge of the countertop by driving screws up through the countertop and into the bottom edge of the backsplash (figure 17-13).
4. Laminating a Countertop
   a. Laminate Trimming Tools and Methods
      - Pieces of laminate are first cut to a rough size.
      - Strip is then cemented to the edge of the countertop.
      - Its edges are flush-trimmed.
      - Laminate is then cemented to the top surface.
      - The overhand is then bevel-trimmed even with the laminated edge (figure 17-14).
   b. Cutting Laminate to Rough Sizes
      - On a table saw.
      - By guiding a laminate trimmer with a flush trimming bit along the straightedge.
   c. Using Contact Cement
      - The bond is made on contact without the need of clamps.
      - On porous material, a second coat is required after the first coat dries.
      - The cement is dry if no cement remains on the finger.

- If contact cement dries too long it will not bond properly.
- Pressure must be applied to the entire surface (figure 17-15).

d. Laminating the Countertop Edges

- Apply coats of cement to the countertop edges and the back of the edge laminate.
- Apply the laminate to the front edge of the countertop (figure 17-16) so the bottom edge, top edge, and ends overhang.

e. Trimming Laminated Edges

- If laminate has been applied to cover the edges of another laminate, a bevel trimming bit must be used to trim the overhanging ends.
- Ball bearing trimming bits have live pilots.
- When using a trimming bit with a dead pilot, the laminate must be lubricated where the pilot will ride.
- Use a file to smooth the top edge flush with the top surface.

f. Laminating the Countertop Surface

- Apply contact bond cement to the countertop and the back side of the laminate. Let dry.
- Place thin strips of wood about a foot apart on the surface.
- Lay the laminate to be bonded on the strips or slats (figure 17-18).
- Make contact on one end. Gradually remove the slats one by one until all are removed.
- Trim the overhanging back edge (figure 17-19).

g. Laminate Seams

- Tight joints can be made by clamping the two pieces of laminate in a straight line on a strip of 3/4 inch stock. Butt the ends together or leave a space less than 1/4 inch between.
- Using one of the strips as a guide, run laminate trimmer, with a flush trimming bit installed (figure 17-20).

h. Laminating Backsplashes

- Backsplashes are laminated in the same manner as countertops.
- Use a little caulking compound between the backsplash and countertop (figure 17-21).

i. Laminating Round Corners

- Strips of laminate can be cold bent to a minimum radius of about 6 inches.
- Heating the laminate to 325 degrees Fahrenheit uniformly over the entire bend will facilitate bending to a minimum radius of about 2 1/2 inches (figure 17-22).

D. Cabinet Doors

1. Types of Doors

a. Overlay Doors

- The overlay type of door laps the entire thickness of the door over the opening.
- Easy to install as it does not require fitting in the opening.
- European-style cabinets feature doors that completely overlay the front edges of the cabinet (figure 17-24).

b. Lipped Doors

- The lipped door has rabbeted edges that overlap the opening by about 3/8 inch on all sides.

c. Flush Type
- The flush type door fits into and flush with the face of the opening.

2. Door Construction
   a. Doors are classified as solid or paneled.
   b. Both solid doors and paneled doors may be hinged in overlay, lipped, or flush fashion.

E. Hinges

1. Surface Hinges
   a. Surface hinges are applied to the exterior surface of the door and frame.
   b. May be offset for lipped doors (figure 17-27).
   c. Used when it is desired to expose the hardware.
2. Offset Hinges
   a. Offset hinges are used on lipped doors.
   b. Semi-concealed hinge.
   c. Concealed offset hinge (figure 17-28).
3. Overlay Hinges
   a. Overlay hinges are available in semi-concealed and concealed types.
   b. Concealed overlay hinges are made for a specific amount of overlay, such as 1/4, 5/16, 3/8, and 1/2 inch.
4. Pivot Hinges
   a. Usually used on overlay doors.
   b. Used when there is no face frame and the door completely covers the face of the case (figure 17-30).
5. Butt Hinges
   a. Used on flush doors when it is desirable to conceal most of the hardware.
   b. They are not often used on cabinets because they take more time to install than other types (figure 17-31).

F. Hanging Cabinet Doors

1. Surface Hinges
   a. Apply the hinges to the door.
   b. Shim the door in the opening so that an even joint is obtained all around.
2. Semi-concealed Hinges
   a. Screw the hinges to the back of the door.
   b. Center the door in the opening.
   c. Fasten the hinges to the face frame.
   d. When more than one door is to be installed side by side, rest the doors on the straightedge to keep them in line (figure 17-32).
3. Concealed Hinges
   a. First screw the hinges on the door.
   b. Center the door in the opening.
   c. Tap on the hinge opposite the face frame.
   d. Place the projections of the hinges into the indentations.
   e. Screw the hinges to the face frame.
4. Butt Hinges
   a. Hanging flush cabinet doors with butt hinges is done in the same manner as hanging entrance doors.

5. Installing Pulls and Knobs
   a. They are made of metal, plastic, wood, porcelain, or other material (figure 17-33).
   b. Pulls and knobs are installed by drilling holes through the door.
   c. Then fasten pulls or knobs with machine screws from the inside.
6. Door Catches
   a. Doors without self-closing hinges need catches to hold them closed (figure 17-35).
   b. Magnetic catches are widely used.
   c. Friction catches are installed in a similar manner to that used for magnetic catches.

G. Drawer Construction

1. General
   a. Drawers are classified as overlay, lipped, and flush in the same way as doors.
   b. Drawer fronts are generally made from the same material as the cabinet doors.
   c. Drawer sides and backs are generally 1/2 inch thick.
   d. The drawer bottom is usually made of 1/4-inch plywood or hardboard.
2. Drawer Joints
   a. Typical joints between the front and sides of drawers are the dovetail, lock, and rabbet joints (figure 17-36).
   b. Joints normally used between the sides and back are the dovetail, dado and rabbet, dado, and butt joints.
3. Drawer Bottom Joints
   a. Fitted into a groove on all four sides of the drawer (figure 17-39).
   b. In some cases, the drawer back is made narrower, the four sides assembled, the bottom slipped in the groove, and its back edge fastened to the bottom edge of the drawer back (figure 17-40).
4. Drawer Guides
   a. Some have a single track mounted on the bottom center of the opening.
   b. Others may be centered above or on each side of the drawer (figure 17-41).

H. Procedures

1. Installing Manufactured Cabinets
   a. Cabinet Layout Lines
      - Measure 34 1/2 inches up the wall.
      - Draw a level line to indicate the tops of the base cabinets.
      - Measure and mark another level line on the wall 54 inches from the floor. Next mark the stud locations of the framed wall (figure A).
   b. Installing Wall Units
      - Start the installation of wall cabinets in a corner.
      - Drill shank holes for mounting screws through mounting rails usually installed at the top and bottom of the cabinet.
      - Fasten the cabinet in place with mounting screws of sufficient length to hold the cabinet secure.
      - Do not fully tighten the screws (figure A).
      - The next cabinet is installed in the same manner.
      - Align the adjoining stiles so their faces are flush with each other.

- Clamp them together with C clamps.
- Screw the stiles tightly together (figures B and C).

c. Installing Base Cabinets

- Start the installation of base cabinets in a corner.
- Shim the bottom until the cabinet top is on the layout line.
- Level and shim the cabinet from back to front.
- Adjust the pencil dividers so the distance between the points is equal to the amount the top of the unit is above the layout line.
- Scribe this amount on the bottom edge of the cabinets by running the dividers along the floor (figure A).
- Cut both ends and toeboard to the scribed lines. Replace the cabinet in position.
- Fasten it loosely to the wall. Align and clamp the stiles of adjoining cabinets. Fasten them together.
- Finally, fasten all units securely to the wall (figure B).

d. Installing Countertops

- After the base units are fastened in position, the countertop is cut to length.
- Use pencil dividers to scribe a line on the top edge of the backsplash.
- Plane or belt sand to the scribed line.
- Fasten the countertop to the base cabinets with screws up through triangular blocks usually installed in the top corners of base units.
- Sink cutouts are made by carefully outlining the cutout and cutting with a saber saw or router.

Answers to Text Questons

Chapter 1

Hand Tools

1. d	4. d	7. d	10. c	13. b
2. d	5. d	8. b	11. c	14. c
3. b	6. a	9. b	12. a	

Chapter 2

Power Tools

1. d	5. a	9. b	13. b	17. c
2. d	6. a	10. d	14. a	18. a
3. c	7. d	11. c	15. a	19. b
4. b	8. d	12. a	16. d	20. d

Chapter 3

Wood and Wood Products

1. d	4. b	7. a	10. c	13. c
2. c	5. c	8. a	11. d	14. d
3. c	6. c	9. a	12. d	15. b

Chapter 4

Fasteners

1. a	4. c	7. b	10. d	12. c
2. c	5. b	8. d	11. c	13. c
3. d	6. c	9. b		

Chapter 5

Blueprints, Codes, and Building Layout

1. c	5. b	9. d	13. d
2. d	6. a	10. c	14. c
3. a	7. c	11. b	15. b
4. b	8. b	12. b	16. b

Chapter 6

Concrete Form Construction

1. d	5. b	9. b	13. d	17. c
2. a	6. b	10. d	14. d	18. d
3. c	7. d	11. d	15. a	19. d
4. d	8. d	12. a	16. b	20. a

Chapter 7

Floor Framing

1. c	4. d	7. d	10. d	13. c
2. b	5. b	8. b	11. d	14. a
3. d	6. c	9. b	12. b	15. a

Chapter 8

Wall and Ceiling Framing

1. b	5. b	9. b	13. a	17. b
2. a	6. b	10. b	14. d	18. a
3. c	7. c	11. d	15. a	19. b
4. a	8. d	12. c	16. a	20. d

Chapter 9

Scaffolds, Ladders, and Sawhorses

1. c	4. a	7. b	10. d	13. a
2. d	5. b	8. a	11. d	14. d
3. c	6. d	9. c	12. a	15. b

Chapter 10

Roof Framing

1. a	5. c	9. c	13. c	17. d
2. a	6. b	10. a	14. b	18. a
3. d	7. c	11. a	15. b	19. d
4. c	8. c	12. a	16. a	20. d

Chapter 11

Windows and Doors

1. c 5. b 9. c 13. a 17. d
2. d 6. a 10. c 14. b 18. d
3. c 7. a 11. c 15. d 19. b
4. b 8. b 12. a 16. b 20. d

Chapter 12

Roofing

1. b 4. d 7. a 9. a 11. c
2. d 5. b 8. b 10. a 12. d
3. a 6. d

Chapter 13

Siding and Decks

1. d 5. c 9. b 13. d
2. d 6. b 10. a 14. c
3. c 7. b 11. c 15. a
4. b 8. c 12. b 16. c

Chapter 14

Insulations and Wall Finish

1. d 4. d 7. c 10. a 13. b
2. a 5. a 8. b 11. b 14. a
3. a 6. d 9. d 12. b 15. d

Chapter 15

Interior Finish

1. d 5. c 9. c 13. c 17. c
2. d 6. c 10. c 14. c 18. c
3. c 7. b 11. a 15. a 19. c
4. d 8. b 12. b 16. d 20. d

Chapter 16

Stair Framing and Finish

1. b 5. c 9. a 13. b 16. b
2. c 6. c 10. a 14. b 17. c
3. c 7. c 11. d 15. a 18. b
4. c 8. b 12. b

Chapter 17

Cabinets and Countertops

1. b 4. c 7. b 10. a 13. b
2. b 5. b 8. c 11. a 14. b
3. a 6. d 9. c 12. d 15. c

Power Tools Assessment Sheet

Under the supervision of your instructor and observing all safety regulations and rules, you will perform the following power tool operations. When you can perform each of these operations correctly and understand all rules about each tool, initial the form. Your instructor will initial each operation when he/she is satisfied with your competence. When he/she is satisfied that you are safe and knowledgeable on that tool, you will be allowed to use that tool to work in the H.B.I. shop or job site.

	Date	Instructor's Initials	Student's Initials
Circular Saw			
Install saw blade and adjust for depth and angle	____________	________	________
Cross cuts	____________	________	________
Cross cutting with a bevel	____________	________	________
Steep angle cuts	____________	________	________
Steep angle cuts with bevel	____________	________	________
Cut sheet goods	____________	________	________
Plunge cuts	____________	________	________
Cut with rip guide	____________	________	________
Saber (Jig) Saw			
Install a saw blade	____________	________	________
Adjust the base for an angle	____________	________	________
Make a cut and cut shape	____________	________	________
Make a plunge cut	____________	________	________
Reciprocating Saw (Sawzall)			
Install blade	____________	________	________
Make a cut	____________	________	________
Electric Drill			
Install a twist drill bit and drill a hole	____________	________	________
Install a spade bit and drill a hole	____________	________	________
Install an auger bit and drill a hole	____________	________	________
Install a hole saw and drill a hole	____________	________	________
Screw Gun			
Remove and reinstall PT2 tip	____________	________	________
Adjust head to proper depth for drywall	____________	________	________
Install 100 1/4" and 10 1/2" screws flush	____________	________	________

	Date	Instructor's Initials	Student's Initials
Power Planer			
Review parts of planer	________	______	______
Make thickness setting	________	______	______
Plane a surface	________	______	______
Router			
Install bit and collet and adjust depth	________	______	______
Set up and rout a dado	________	______	______
Set up and rout a rabbet	________	______	______
Set up and rout with an edge-guided bit	________	______	______
Use edge guide	________	______	______
Belt Sander			
Remove and install a sanding belt	________	______	______
Adjust belt tension	________	______	______
Sand a surface	________	______	______
Clean belt cleaner	________	______	______
Finish Sander			
Change paper	________	______	______
Sand block wood to finish readiness	________	______	______
Pneumatic Nailers			
Load fasteners in magazine	________	______	______
Make a trial fastening	________	______	______
Adjust air pressure as needed	________	______	______
Single fire and rapid fire fastening	________	______	______
Powder-Actuated Drivers			
Review safety protocol of device	________	______	______
Select and load fastener	________	______	______
Select and load charge	________	______	______
Position material and driver for fastening	________	______	______
Pull trigger and reload for next fastening	________	______	______
Table Saw			
Adjust fence	________	______	______
Lockout and change saw blade	________	______	______
Install splitter	________	______	______
Rip and bevel a board	________	______	______
Cross cut a board	________	______	______
Set up and cut a dado and rabbet	________	______	______

	Date	Instructor's Initials	Student's Initials
Miter Saw			
Make a straight cut	______	______	______
Cut miter	______	______	______
Cut a miter to long point	______	______	______
Cut a miter to short point	______	______	______
Cut a bevel	______	______	______
Cut a compound miter	______	______	______
Band Saw			
Adjust upper guide to proper height	______	______	______
Cut stock using a miter guide	______	______	______
Cut stock using a rip fence	______	______	______
Cut stock free hand following a curved line	______	______	______
Jointer			
Adjust fence and infeed table	______	______	______
Joint an edge	______	______	______
Joint a taper	______	______	______
Drill Press			
Adjust table level	______	______	______
Set up chuck and check	______	______	______
operation	______	______	______
Drill a hole	______	______	______
Shaper Table			
Assemble shaper	______	______	______
Install bit	______	______	______
Cut rabbet	______	______	______
Cut dado	______	______	______
Make 1/4 round	______	______	______
Combination Disk/Belt Sander			
Square disc platform	______	______	______
Change disc	______	______	______
Adjust belt	______	______	______
Sand a piece of wood	______	______	______
Extension Power Cords and GFCI			
Inspect power cord	______	______	______
Correctly connect power tool to extension power cord and GFCI	______	______	______
Inspect GFCI	______	______	______

	Date	Instructor's Initials	Student's Initials
Stepladder			
Inspect stepladder	________	______	______
Correctly set up stepladder (all sizes)	________	______	______
Climb and dismount stepladder (all sizes)	________	______	______
Carry a stepladder (all sizes)	________	______	______

SECTION TWO

Note to Instructors

Note to Instructors

The area of greatest problem for students tends to be math. Basic math worksheets were developed to help students become more comfortable using formulas. Encourage the use of calculators for math accuracy. The student *must* understand the math concept prior to total dependency on a machine. The first sheets should be done longhand. The introduction of calculators will begin after the review is completed. Whenever possible, putting math problems into real-life situations makes learning much easier. You may find some of your students too advanced for the beginning math. Do not bore them with needless exercises. As individuals, any attempt to clone a class ends in disaster. Bright students quickly lose interest and students who need help become too frustrated. Diversity of students' needs and abilities can both challenge and, at times, overwhelm an instructor.

To be an effective trainer, one has to be aware of how people learn. A person generally has a dominant style of learning. The instructor must have an understanding of how people learn through their senses so that information and skills can be effectively presented.

Learning Styles

The ways that students learn have been researched extensively; subsequently, there are many theories about learning styles. The focus here will be on three learning styles. The three senses used most often in training are: seeing (visual), hearing (auditory), and touching or doing something (kinesthetic).

The Visual Learner

Most learners have a dominant visual learning style. The saying, "a picture is worth a thousand words," describes the use of visualization very well. An instructor who uses visual aids to present information will be more successful than one who relies wholly on lecture. Information is presented in real or conceptual form. For example, when describing the concept of wood warping, the instructor may show a piece of twisted wood. The cause would require a more detailed explanation, as would preventative measures. Many of the operations performed in the construction industry need to be visually presented for a clearer understanding. The object of the lesson and its content will dictate the type of presentation used. For example, if knot tying were the topic, a lecture about which end goes over what part of the rope would be very confusing. A demonstration followed by students actually tying pieces of rope would be the winning combination. Although demonstrations are the most widely used form of presentation in apprenticeship training, there are many times when a videotape or overhead slides may be better. A video on framing could move an entire building into your classroom. A week of work can be presented in minutes. Areas of interest can be replayed for clarity. Eventually, the student would get hands-on experience, but knowing the sequence of construction will lessen confusion.

A visual learner is determined by the interest he or she shows in looking at pictures, diagrams, videos, and illustrations. Visual learners might doodle in their notebook and would probably draw you a map instead of telling you how to get somewhere. Students who are in the front of the group so they can see the action are visual learners. When giving directions, the instructor should write them on the board or on paper, as well as verbally announcing them. A sketch will help a student who is having difficulty understanding something.

The Auditory Learner

We hear two different types of auditory signals: noise and verbal communication. Both are very important and needed for learning. A binding saw makes a noise that instantly draws the attention of the instructor. On the job we become accustomed to the normal work noises and are able to identify many problems by the sounds they make.

Verbal communication is the interaction between two or more people. Communication is a two-way process in which information is *sent* and *received*. In education, what we send needs to be received and understood for learning to take place. Communication is made up of several components. The words used to convey the intended message are referred to as verbal content. These words have meaning.

As important as verbal content is, students show it accounts for only 7 percent of any communicated message.

The other components are inflection and body language. Together they make up the remaining 93 percent of the message you send. Inflection is the way that your voice is used to emphasize parts of the message. It includes voice tone, volume, and rate of speech. Speaking very rapidly makes a different impression than someone saying the same thing in a slow and deliberate manner. Inflection is about 38 percent of the meaning of the message. Body language includes your gestures, eye contact, facial expressions, and body posture. At 55 percent of an oral message, body language can often express more than words.

Receiving the message is just as important as sending it. Most people could be much better listeners than they are. Good listening skills need to be developed. Listening carefully does not always mean you will get the intended message. Many messages and directions are vague and confusing. For effective communication, both parties must understand each other.

The auditory learner will listen to directions rather than needing them written down. To aid the student in learning, project your voice so that any secondary noise does not interfere with the message. Keep your instructions clear and concise. Articulate and speak slowly so that nothing is missed.

The Kinesthetic Learner

This style of learner uses all of the senses, with the center around manipulative skills. Being physically involved in the learning environment, the kinesthetic learner uses sight, sound, smell, and touch simultaneously. This is the preferred mode to teach technical and skill development. Scaled-down models, samples, and multiple practice pieces that can be touched by the learner will enhance the learning experience.

The cost of materials and supplies can directly affect the ability of the instructor to provide a challenging learning environment. The amount of hands-on training will be limited. Keep in close contact with vendors, contractors, manufacturers, distributors, and government agencies that may offer discontinued or surplus items for training. Providing realism to the learner will add to the credibility of the instructor.

Safety is the most important environmental factor. A student must show competency at one level before progressing to the next skill. Kinesthetic learners are the people who must have their hands on everying. *Instructors teach in their preferred learning style.*

Retention Rate

The chart below shows the average retention rate. We remember little of what we hear, and not much more of what we read. By introducing an audio/visual format into the presentation, retention doubles. Another 10 percent is added with a demonstration. Students will remember half of the information after participating in a group discussion. When hands-on practice is included, the rate jumps to 75 percent. Allowing a student to teach another immediately after learning helps ensure that the concept or skill has been mastered. The instructor should oversee the exchange of information to alleviate any mixed or erroneous messages.

LECTURE	5%
READING	10%
AUDIO/VISUAL	20%
DEMONSTRATION	30%
GROUP DISCUSSION	50%
PRACTICE BY DOING	75%
TEACH OTHERS	90%
IMMEDIATE USE OF LEARNING	

AVERAGE RETENTION RATES

The need to keep the classroom multisensory and stimulating is obvious. This can be done in a number of ways. Since "a picture is worth a thousand words," use posters, pictures, diagrams, and sketches to help make your point. Subscribe to a variety of magazines such as *Fine Woodworking, Work Bench* and *Fine Homebuilding*. If budget constraints do not allow a number of subscriptions, try pooling resources with other instructors. Each instructor subscribes to one publication and agrees to pass it on to the others when they are done. An April copy read in June still has the information that is made even better because it is free. The exchange can be between departments, within the school district, or in cooperation with the library. Videos can be shared in a like manner. Over time, you will accumulate a ready resource library. This can be done within a school or department.

Sources for videos are open for the imagination. Manufacturers have tapes about their products to ensure proper installation and to promote their use. Weyerhauser has a library in its corporate headquarters in Washington State. Videos are loaned free of charge, but you must pay for the UPS shipping and insurance to return them. The tapes are helpful in showing the students how trees are harvested and the many procedures in the mill. You can purchase any number of how-to tapes through magazines or at local lumber stores. Some stores have the tapes available to rent. This allows you to use it for a day and return it after that segment of the class is over. This is also an inexpensive way to preview the film prior to buying it.

There are entire series developed and available. The Public Broadcasting Station (PBS) show *Hometime,* with hosts Dean Johnson and Joanne Liebler, have dozens of step-by-step instructional videos. They are moderately priced due to the contributions of Chevy Trucks. There are mini-commercials slipped into the tapes, but the contents remain good. Each tape comes with a brief project guide designed for do-it-yourself home repairs.

Your choice of videos is not limited to purchasing. Obtain permission to record and use television shows that are relevant to your class. PBS seems to provide the best selection with *New Yankee Workshop, Woodcarving with Rick, This Old House,* and *Hometime.* Contact your local station for information about copyrights to replay their programs. Viewing these shows with the students can prompt a group discussion about technique, materials, tools, and safety.

Contact manufacturers for speakers to demonstrate and bring samples of their products. Most sales and public relations people are proud of their products and are eager to put them in your shop. The students in training today will be on the job tomorrow. They will continue to use the tools, materials, and procedures they are most familiar with. Give the sales department the opportunity to put their product in the students' hands. Some companies will provide their systems at reduced, or no, cost to ensure familiarity with their product. Good communications with suppliers can help keep you abreast of new developments in the field. Vendors or manufacturers sometimes have *nuisance* materials that are not cost-effective to keep on inventory.

Books provide the student with volumes of knowledge at their fingertips. Besides the usual channels to acquire books — publishers, magazines, marketing, and so on — a great resource is used book stores. They have specialty, how-to, and vintage books for a fraction of their original retail value. Books offer the student the chance to help themselves discover the answers to their questions. Their quest for knowledge often exceeds the instructor's lesson plan. Anything instructors do to stimulate their students improves the quality of the craftspeople they become.

Multisensory Classroom

Section 1 of the text can be supplmented with a variety of activities, both in- and outside the classroom. Much of the valuable information found in these chapters can be made more palatable by having the students use their senses to touch, hear, smell, and see the different properties of wood.

Class Project

A class project gives the students the opportunity to put it all together. This allows their skills and talents to be recognized beyond the confines of the classroom. What you choose to do is limited by your imagination, the program's length, student diversity, community involvement, local and jurisdictional laws, supplier generosity, and always, the *budget.* Find a need in the community and fill it. Word of mouth advertising of your program will stimulate community interest at school.

Class projects could be small, useful items such as a laminated tote for garden tools, a one-step bench for reaching the top shelf, or racks for many items, from CDs, wine, and videos to ties. Children's toys — trucks, trains, wooden rocking animals (horse or elephant), puzzles, or even blocks of wood painted with bright primary colors (nontoxic, of course) — are inexpensive ways to recycle many wood scraps. These items can be given to shelters for battered women and children, retirement homes, and as holiday gifts for underprivileged families. Contractors and suppliers will often have surplus materials to donate. Students will benefit both in the heart and hands as they increase their skills using tools while helping others.

Another group of items to consider for class projects is outdoor necessities. Redwood picnic tables, a child's playhouse, fort, or exercise center, a gazebo, planters, dog houses (use kind plans included), or a backyard storage shed are possibilities for projects. These can be raffled during an open house or other school function.

If drafting is a part of your program, the students could design and build a hot dog stand/trailer. This would require the combined effort of many departments. The welding shop (the frame), auto shop (undercarriage and wheels), carpentry (framing and finish), plumbers, and electricians are all needed for this undertaking. Once completed, the trailer could be licensed for the road and sold or used by the school. Note that this could also be done as a custom order, using the customer's requirements.

Have a house design contest with area architects and contractors as judges. The winning design could be built and auctioned off, or the students could each build their homes at 1/12 scale. A 1200-square-foot rambler would only need a 30-inch by 40-inch area in the shop. Styrofoam can be used for the foundation and a glue gun will hold the framing together. Creativity should be encouraged during planning and finish details. The framing of roof rafters or trusses and laying out stair stringers will be very challenging, with a small margin of error.

Your class can volunteer labor for nonprofit organizations in the area. They could help build a house for Habitat for Humanity, build a wheelchair access ramp at someone's home, or get involved with a service group to help low-income homeowners maintain their homes. The YMCA or YWCA, the Boys' and Girls' Clubs, parks and forest departments, local labor unions, and church groups are a few of the groups to contact to help the community. If building an entire house is too much time away from the classroom, consider doing just the portion that is related to their studies. From sidewalks to roofing, nothing beats the field experience they receive.

NOTE: All of the suggested activities and projects need to be approved by the governing advisory board or school board.

Ergonomics

"Ergonomics" is a term derived from the Greek words *ergon,* meaning work, and *nomos,* meaning laws. Ergonomics is that field of study that is concerned with the problems and processes involved in designing things for effective human use, and in creating environments that are suitable for human living and work. It recognizes that work organization, work methods, equipment, facilities, and tool design all influence the worker's productivity, motivation, fatigue, and likelihood of sustaining an occupational injury or illness."

Construction is one of the most hazardous occupations in the country. Workers are killed or injured by falls, electrocutions, falling objects, cave-ins, crane accidents, and even bad housekeeping. Many people in the trades work and live in pain — in their backs, hands, shoulders, and knees. Others who suffer this pain can no longer work.

These problems are now recognized as Cumulative Trauma Disorders (CTDs), disorders that are caused by the daily stresses and strains of work. These musculoskeletal injuries and strains develop undetected until pain or dysfunction disrupts the ability to work and the quality of life. Injured backs, carpal tunnel syndrome, and knee and shoulder problems do not have to be part of the job. These diseases are preventable.

The first step in prevention is to understand how these injuries occur. The shoulders, elbows, wrists, hands, fingers, knees, and back are mainly affected by CTD. Work activities performed each day can cause tiny injuries (traumas) to these parts of the body. The trauma is so small, it will go undetected until all of the small traumas are added up and there is pain. CTDs may take weeks, months, or even years to develop. How work is done today will affect the health of the carpenter tomorrow. Almost 60 percent of all work-related illness are CTDs.

Ergonomics can be used to redesign the way work is done. The work will still be done, oftentimes at a quicker rate. Changes may be as simple as storing materials at waist height instead of on the floor to reduce bending. More complex changes will involve redesigning tools or altering the work process itself. Manufacturing of tools not yet drawn up is in the future, but changes in the workplace can begin today. People are sometimes resistant to change because they cannot see the value in doing things differently. If our students can be trained with an open mind as to how many tasks are done, they may be spared pain later in life.

Muscles

Muscles make it possible to move. Tendons attach muscle to the bone. Where two bones meet, there is a joint. A ligament stretches across the joint to connect the two bones. The ends of the bones have cartilage to protect the bones from grinding against each other. There is lubricating fluid between the bone end and the cartilage. Nerves carry messages between the brain and body parts. Blood vessels bring oxygen and nutrients from the heart to the muscles. Waste products are removed when the blood returns to the lungs for more oxygen.

Most of these body parts can be observed in action by looking at the back of the hand. The bluish arteries supply the hand with blood. By wiggling the fingers, the tendons will jump rapidly across the hand. The smooth movement of the fingers is made possible by the internal lubricating system. Making a fist will show the muscles at work.

Muscles can only contract (shorten) or relax (lengthen). Because they work one way, they are usually found in pairs. Moving muscles helps pump the blood into the tissue. Moving muscles are performing dynamic activity. When using the hand saw, the arm muscles are doing dynamic activity. The opposite is static activity. *Static* means motionless. The arm used to hold the wood when sawing is in a static state. During static activity the muscles remain contracted. They do not relax or get enough blood. The muscles get tired very quickly. This is called *fatigue.* Muscles can recover from work if they are allowed to rest. When they are not given enough time to recover, they will ache.

Running will produce whole body fatigue. Using a group of muscles is local muscle fatigue.

When nailing overhead, the shoulder muscles work to keep the arm up. This will tire the shoulder muscles. Localized muscle fatigue can also happen when small muscles are used to do a job that they are not strong enough to do. Pinching the top of a sheet of 3/4-inch plywood to move it will quickly fatigue the hand. But using the whole arm to support weight with the hand wrapped around the bottom will greatly increase the amount of work that can be done.

Logically, the smaller muscles will get tired faster than large muscles. More force is needed to use smaller muscles because they have less strength. A pinch grip uses the small hand muscles to hold an object with the tips of the fingers. An example would be holding a nail by the head between the thumb and forefinger. A power grip uses the larger arm muscles to hold an object with the whole hand. A handshake is a power grip, as is the grip used to hold onto the hammer handle when nailing.

Fatigue causes pain and discomfort. The body can usually recover with rest. If pain is present daily, the body may not be getting enough recovery time. Resting the main muscles that are working will help recovery. A micro rest will help. Using different muscles or pausing for a few seconds is a micro break. This is better for the muscles than one or two longer periods of rest after continued work. The parts of the body can suffer lasting damage without recovery. Getting enough nighttime rest is a valuable part of the body's recovery.

Tendons, Nerves, and Blood Vessels

Bones and muscles are harder than tendons, nerves, and blood vessels. When they are in contact with bones and muscles, these soft tissues can be trapped or flattened, stretched or squeezed. When a muscle is contracted, tendons are pulled and stretched. Inflammation can occur with repeated use. Inflammation is the body's way of stopping work to allow the injured part to heal. Swollen tendons cannot function properly. The ability of the muscle to work is reduced and there may be pressure against nerves and blood vessels.

Nerves that are pinched cannot relay messages properly. The result might be numbness, tingling, or a burning feeling. Nerves may be squeezed, compressed or trapped by bones, muscles, inflamed tendons, or damaged by handles or edges that press against the palm of the hand.

Blood vessels can be damaged in the same way as nerves. Tissues that do not receive a blood supply will die. Muscles will weaken and there will be a loss of feeling beyond the pinch point. Fingertips will begin to turn white when the blood supply is restricted. Flexing will help stimulate the blood flow and the fingers will quickly turn red.

What Can Be Done?

The muscle effort needed to do the job can be reduced by using larger muscles (changing the grip), repositioning the body to escape an awkward posture, choosing a tool that keeps body parts in neutral positions, and using gloves to avoid direct contact with cold or slippery tools.

Students can better understand how the position of the body affects its ability to do work by having them participate in the following exercise.

1. Working in pairs, have one person hold a pencil in a straight-wrist power grip (the arm is parallel with the floor and the pencil will be plumb).
2. The partner tries to remove the pencil by pulling up.
3. Repeat with the wrist bent downward. (The pencil is now parallel with the floor.) **It should be easier to pull the pencil out of the grip when the wrist is bent.**
4. Have the partners switch and repeat steps 1–3.

Discuss the effort needed to hold the pencil in the bent wrist position. When the body is in an awkward position, greater effort is needed to perform a task that would not require much effort using a neutral posture.

The duration of an activity adds to the risk factors of developing a CTD. Obviously, the longer an activity is done, the greater the risk. Dividing the activity into segments or simply shifting positions will shorten the exposure. Using muscles in a static position will cause them to tire very quickly. Allowing dynamic work or a mixture of the two is better. The following exercise helps demonstrate how this is done.

This is not a competition. Pay close attention to how your arms feel.

1. Working in pairs or small groups, use two 5-pound weights.
2. Have a person hold a weight in each hand.
3. Holding the arms straight out from the body, keep one arm up. The other arm is lowered every 10 seconds.
4. The students without weights are to watch the static arm and make sure it does not creep downward.
5. Time is kept until the static arm starts to hurt.
6. Change roles.

Discuss the effects of static versus dynamic work. The contracted muscles cut off the blood supply and cause early fatigue. The dynamic arm was alternating contraction and relaxation. This helped pump the blood to bring oxygen and get rid of waste products. The muscles were able to continue to work longer after the static muscles could not. Microbreaks for the working muscles can be a few second pause or changing the work briefly. Vary the tasks as much as possible, which also reduces the effects of repetition.

The objective of this area of training is to teach the students to work smarter, not harder. This is not a message to work less or be afraid of honest hard work. Quality work can be performed in larger quantities by healthy workers who are not distracted by constant pain. Adjusting the work height by moving the materials onto saw horses with a forklift takes only minutes to do, but the worker's back is spared a whole day of bending and lifting. Working off of a ladder to drill holes at waist height, rather than standing on the ground to drill at shoulder height, will make the holes easier to drill and keep the muscles fresh for the next assignment.

Additional information on this subject is available through the Department of Labor and Industry or your local OSHA office.

Back and Back Pain

The improvements in knowledge and technology continue to make the workplace safer for the worker. Keeping abreast of these changes can make the difference between injury or safety. The construction trades have a higher rate of back injuries and muscle strains than many other occupations. The nature of the work puts physical stress on the body. Studies show that one-quarter of all construction injuries include the back.

Back pain and discomfort of the joints is a major factor in forced retirement from the trades. Many people who change to less demanding occupations often experience a dramatic reduction in income.

Only 23 percent of back injuries are caused by sudden accidents like falling, tripping, or slipping. Many of these could have been avoided by improved housekeeping. Regular cleanup of waste material and construction debris in the work area greatly reduces risks. Proper material storage for ease of access also improves productivity with fewer wasted motions.

Most back injuries often start without warning. They are the result of wear and tear encountered daily in the field. Using the body improperly with repeated twisting, awkward postures, and heavy lifting takes it toll. Ligaments can be painlessly overstretched, disks overcompressed, and muscle imbalances created that will set the body up for pain later. Once back pain is experienced, the chances of recurring pain are greatly increased. Nearly two-thirds of all back injuries are caused by abuse of the worker's body. Through education and training, back injuries can be reduced. Simple routines that minimize the strain on the back and improve posture, when coupled with an understanding of how the back works and training in lifting techniques, are effective. Prevention is the key to success.

Spinal Column and Muscles

Counting the tail bone, the back consists of 25 cylinder-shaped vertebrae. This vertical stack of hollow bones is held together with straps called ligaments. The nerves run up the spine through the holes in the disks, like communications wiring in a conduit. Nerves carry messages from the brain to the specific parts of the body and return information to the brain about the conditions throughout the body. At different levels, the nerves leave the protective cover of the spine and pass between the vertebrae to reach their destination in the body. This fragile yet resilient column supports the weight of the body (dead load) and the additional weight of what is carried or lifted (live load) without pressing on the nerves. Small spinal joints called *facet joints* link the vertebrae to each other.

The bones are separated by disks that act as hydraulic shock absorbers. The invertebrate disk is made up of two parts. The soft gel-like center is surrounded by a hard fibrous ring. Each disk is attached to the vertebrae above and below it. There is no such thing as a *slipped disk;* a disk cannot slip out of place. Disks can actually withstand much wear and tear. They are continually being compressed by the effects of gravity. If the outer ring is damaged, part of the gel can leak out. This causes a bulge that presses on the nerves. The pain may be felt in the leg or other places as well as the back. With age, the disks lose water (disks are composed largely of water). This is a normal part of the aging process that is accelerated by poor back care. Carrying uneven loads puts added stress on the disk, as does twisting while lifting or carrying a load.

Back muscles are attached to the vertebrae and run up and down connecting to body parts on the front, sides, and back. Movement of the body is controlled by the contraction (tightening) and relaxation of muscles. It addition, rope-like ligaments join bone ends to support and strengthen joints and prevent abnormal movement. When the front muscles (stomach muscles) tighten, the body is bent forward or prevented from bending backwards, such as when driving screws overhead. Tightening of the back muscles causes the body to bend backwards or prevent falling forward, such as when bending over a work table. Muscles on either side of the body pull toward their side or when carrying a load to the side that will prevent toppling over. An example would be that when carrying a sheet of drywall on the right side, the muscles on the left side keep the body upright. The spine is the only bone between the ribs and the hips. Because of the lack of body structure in the lower back, the body depends on these interwoven muscles and strong ligaments across the abdomen and lower back to provide support. Keeping the muscles in good condition is the goal of preventative measures.

The spine has three curves in the neutral position. Proper posture is critical to maintaining the natural curve. Prolonged unnatural positions can lead to stress and pain. Any position held too long is not good for the back. The little aches and pains are a warning to change positions. Using muscles to control posture will minimize the effects daily life has on the spine. Lower back muscles are short, small, and usually very strong since they are working almost continually to keep the body upright. These anti-gravity muscles need the help of the big, broad stomach muscles designed to take some of the workload. Failure to use the stomach muscles will overload the back and cause an imbalance that leaves the spine susceptible to injury.

The muscles of the body are either working or resting. Muscles can work either dynamically or statically. Dynamic work is done by the muscle alternately contracting and relaxing, such as sawing would involve. During static work, the muscle stays constricted without joint movement. Staying in the rigid static position causes muscle pain. Muscles must relax to allow oxygen rich blood to circulate through the cells.

Working Smarter

Learning to work in a position that keeps the spine neutral takes practice. It is critical to do so to allow the spine to carry loads in the way it was designed to carry them. The basic standing position is with legs slightly spread and bent and the lower back straight. This will use the thigh muscles a lot. When done correctly, the fatigue will be there, not in the back. Lifting from this position is relatively safe.

Avoid prolonged working in a bent, crouched, or stooped position. Rearrange your workbench to waist-high to lessen stress. Take regular breaks to stand up and bend backwards to relieve tension. This should be repeated three times for the muscles to relax. Use three forward stretches to relieve tension when overhead work causes an arched position. Learn to use the arms and legs instead of the back to get leverage. Avoid excessive bending of the spine, since parts do wear out. Minimize the wear and tear whenever possible. Remember, small injuries add up to CTDs.

Lifting

Proper lifting starts with looking at the load. Size it up and check for a path that is clear of obstructions. Read any labels or packing data that might state the weight. Get help if needed to ensure safety (pride is not worth a back injury). Use a dolly, pallet jack, or other materials handling equipment when possible. Begin in the basic standing position with one foot slightly forward. Squat with your legs while keeping the back straight. Bending slightly will double the standing disk pressure. Bending over further increases the pressure even more. Get as close to the load as possible and keep it close while transporting. A 10-pound weight increases the pressure on the disks 10 pounds when held close to the body. The same 10-pound load will exert 100 pounds when held at arm's length. Get a good grip with solid contact using the entire palm. Tighten stomach muscles as the lift begins. Let the legs do the work. Lift slowly without jerking. If a turn is needed, change the direction of the entire body by pointing the forward foot in that direction. Do not twist the body. Lower the load slowly, again using the leg muscles to save the back.

Carrying loads with another person requires communication. The lifting strategy must be clear. Lifters should be of similar height to distribute the load evenly. The long load is carried on the same side by both lifters, with the person in front acting as the eyes for both to ensure a safe path. Lift the load from the ground to waist height and then from the waist to the shoulder. When carrying long, flexible loads such as rebar, walk out of step to minimize bounce.

Carrying a load up or down stairs has added dangers. One person is required to walk backwards, since the people are facing each other with the load between them. Use the stomach muscles to support and protect the back. The tallest or strongest person should be at the bottom, where the load is heaviest.

Carrying an uneven load can cause excess stress. Rather than carry a full bucket in one hand, divide the load into two buckets to balance. If a one-handed load cannot be avoided, hold the free arm out or on the hip for balance.

Various types of material handling equipment are available to lift and move loads of different sizes, shapes, and weights. This equipment can save not only time and labor, it can save your back; use them whenever available. A cart with larger wheels moves easier over rough terrain and up stairs than one with small wheels.

Pulleys

The use of a pulley can reduce the effort needed to hoist a load in half. Combining a block and tackle can further reduce the applied force necessary to lift a load.

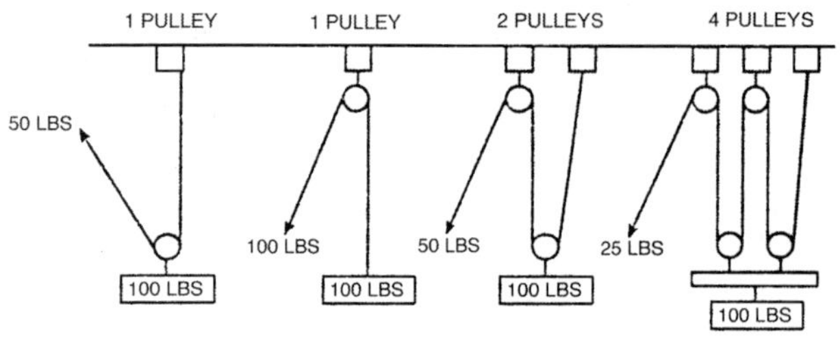

Levers give a mechanical advantage in moving or raising heavy objects. The levers below are used without a fulcrum block. Note that the force is up.

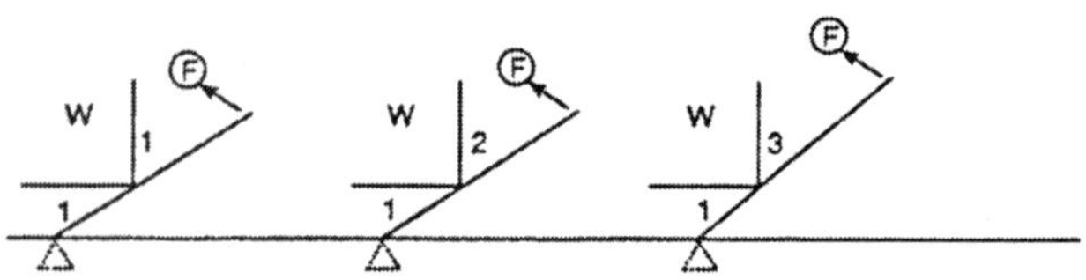

Force = F = 1/2 of Weight W F = 1/3 W F = 1/4 W

The levers below have fulcrum blocks. Note that the force is down.

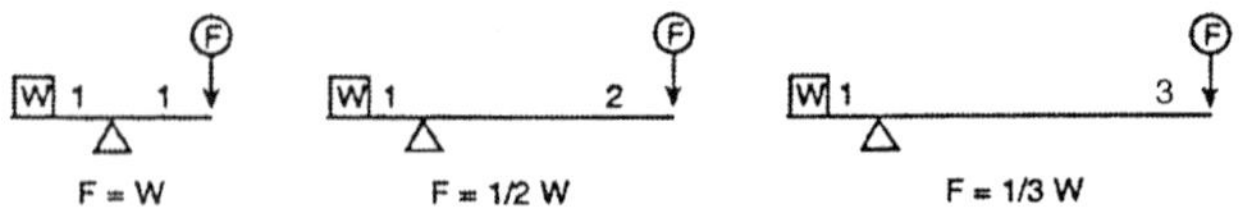

Exercise is a daily activity. Warming up and stretching before work helps the muscles stay in balance. Work tends to overdevelop some muscles, while others, like the abdominal muscles, grow weaker. Effective exercise can be simple and enjoyable.

Hammer Type, Size, and Selection

Hammers come in a variety of weights, from less than 1/2 pound to over 2 pounds. Their handles can be wooden, fiberglass, steel, or any combination of these materials. The grips on the handles can be rubber reinforced to help absorb shock, or have a light coating of vinyl for a better grip. Some new models have a hooked handle, while others simply have small grooves to avoid slippage.

Many studies are under way to help develop the perfect hammer as well as other user-friendly tools. The needs and preferences of the carpenter and the work at hand determine the hammer chosen.

The steel-shanked handle provides the most durability for the repeated stripping encountered in form work. Unfortunately, the impact of driving the nails is transmitted through the handle to the carpenter's arm. The fiberglass handle is also durable and does not transmit as much shock to the arm. As the handle wears, painful slivers become a threat. Many carpenters wrap electrical tape around the handle to protect themselves from harm. The natural fibers in the wooden handle help lessen the trauma to the arm. These handles are made of quality hardwoods such as hickory. To prolong the life of the handle, nails should never be pulled by leaning the hammer to the side in a back and forth motion. The use of a nail bar, wrecking bar, or stripping bar are the best choices when pulling nails. Watch for loose handles on all hammers. A wobbly head could become a flying object and cause injury.

Hammer heads have many different features. Again, the type of work being done contributes to the selection of a hammer. Some carpenters use just one hammer; most own several to suit their needs. The smooth face hammer is used in finish work. To prevent damage to the wood surface, the nails are not driven home with the hammer. The nail set is used to recess the nail head below the surface. When concrete forming is being done, the smooth face does not damage the plywood surface. The claws may be either straight or curved. The weight of the hammer is a personal preference, but the size of the nails and the type of wood will make a difference. Commonly used finish hammers weigh between 16 and 24 ounces.

Corrugated face hammers are used for framing. The irregular surface grabs the nail in a way that makes even a glancing blow productive. When new, the face of these hammers can cause considerable damage to both the wood surface and any skin it may come into contact with. The range of weights for these hammers is 20 to 32 ounces. Many have long handles for added driving power. Some of the heaviest weights have a handle that is shaped like an axe handle for comfort when swinging.

Eye protection must be worn to safeguard against flying nails, wood splinters, and tiny metallic fragments. The development of a consistent swing and rhythm reduces the number of flying nails caused by not hitting the nail squarely on the head. If the carpenter is leading with the wrist, the nail will fly away. Bringing the hammer head down ahead of the wrist will send the nail toward the body.

Ladders and Scaffolds

The use of ladders and scaffolds in the field is unavoidable. The safe use of this equipment is essential. Following basic safety guidelines will reduce the risk of injury.

1. Always place ladder legs on a level, firm surface. Extend brace arm fully and lock into the open position.
2. **Always face the ladder when climbing, keeping three points of contact** (either two hands and a foot or two feet and a hand).
3. Use a ladder that is tall enough for the job — **never stand on the top two rungs of a ladder.**
4. As with all tools, inspect them before using.
5. Do not overreach, as this can cause the ladder to tip over. Get down and move the ladder as your work progresses.
6. Do not leave tools or material on rungs.
7. Assemble scaffolding according to the manufacturer's directions.
8. **Never ride on a rolling scaffold while it is being moved.**
9. Use all safety features required by local or OSHA, including hand-rail, toeboard, approved planks, crossbracing, and outriggers.
10. **Never exceed weight limits.**

Contact your local OSHA office or safety affiliate for videos, slides, films, overhead transparencies, manuals, and/or posters. These are usually offered free of charge, although return postage is needed for some items. Guest speakers can also be arranged through these agencies. Safety cannot be overemphasized. Young people tend to view themselves as invincible and may take more unnecessary risks. Ignorance plays a part in this risk taking. Fortunately, ignorance can be cured — with knowledge.

Material Safety Data Sheets

The carpenter in the field faces numerous dangers; the work is physically hard, tools are potentially dangerous, there are many chemicals, and there is the possibility of a fall. There are many ways to keep the carpenter safe. Being physically fit is a good start. Maintaining a high level of conditioning is essential. This does not mean a carpenter must be built like a linebacker. There are many competent, professional carpenters who are between 110 and 160 pounds. They do not appear as impressive as their 200-pound counterparts, but they tend to work smarter, not harder, to do the same job. The construction field needs more qualified and dedicated women, as well as men.

When working in heavy concrete construction, a carpenter may carry over 60 pounds strapped to his or her body. This added weight comes from boots, fall protection harness, lanyard, tool belt with wrenches, hammer, bars, and other hand tools and bolts. Climbing up and down gang forms that are 20, 40, or 60 feet tall requires strong arms and legs. A small person may have an advantage of less body weight to move. People who are out of condition will have more aches and injuries.

Both hand and power tools present a potential hazard. Any tool left lying where someone could trip on it or kick it down on someone is an easily remedied danger. Physically examining all tools prior to use will also reduce injuries. Electric power tools require special concentration, as the danger of electrocution is present. Misuse of hand tools can cause as many accidents as power tools. Use all tools for the purpose they were intended to perform.

In 1974, the National Institute for Occupational Safety and Health (NIOSH) joined the Occupational Safety and Health Administration (OSHA) in developing a series of complete occupational health standards for substances with existing Permissible Exposure Limits (PEL). NIOSH is responsible for recommended health and safety standards. OSHA has jurisdiction in promulgation and enforcement activities. This joint effect was named the Standards Completion Program. It involved the cooperation of contractors and personnel from NIOSH and OSHA. The NIOSH *Pocket Guide to Chemical Hazards* was developed to make the technical information available to workers, employers, and occupational health professionals. This document is in the public domain and free copies are available from:

Publications Dissemination, DSDTT
National Institute for Occupational Safety
and Health
4676 Columbia Parkway
Cincinnati, Ohio 45226
(513) 533-8287

For information on other occupational safety and health problems, call:

1-800-35-NIOSH

Right to Know

Employees have the right to know all hazards they are being exposed to. A federal law, the Workers Right to Know Act, passed in 1984, requires all states to comply with hazard communication regulations. Standards may vary from state to state, as they are tailored to the industries found there. Manufacturers and importers must make hazard information available through distributors to users in the form of container labels and material safety data sheets (MSDS) if their products contain hazardous chemicals. Employers must provide information and training to their employees regarding any chemical-related hazards they are exposed to. The material safety data sheet checklist was printed by the Washington State Department of Labor and Industries, Division of Industrial Safety and Health.

Safer Job Site

The need to have trained and safety-conscious employees is very great in the construction industry. As contractors work toward providing a safer workplace, workers must also do their part in following company safety policies. Many practices that were once the standard are now foolhardy and illegal by new safety regulations. Most construction companies have developed an Employee Safety Manual, Policy, and Program for their employees and subcontractors to follow. Mandatory attendance by all crafts to weekly job site meetings and strict adherence to the safety policy are expected. Violators of these rules are reprimanded or their employment is terminated. They are an unnecessary risk to the contractor, the project, and to fellow workers. A cooperative effort in the office and the field to continue training and implementation of safety equipment will prove to be mutually beneficial. Employees who know that their employer is looking out for their health and safety are more productive.

MSDS Checklist

	Each item must be provided on every MSDS.	Complete information? Yes	No
1.	Product or chemical Identity used on the label	________	________
2.	Name, address, and phone number for hazard and emergency information	________	________
3.	Date of MSDS preparation	________	________
4.	Chemical and common names of hazardous ingredients **Note:** *This includes all hazardous ingredients which comprise 1 percent or greater of the composition except for chemicals identified as carcinogens, which must be listed if the concentrations are 0.1% or greater. Any chemical ingredient, even if less than 1% of a mixture (or less than 0.1% for a carcinogen), must be listed in the MSDS if there is evidence that the ingredient(s) could be released from the mixture in concentrations which could exceed established "acceptable" exposure limits or if the ingredient(s) could present a health hazard to employees.*	________	________
5.	WISHA permissible exposure limit (PEL), American Conference of Governmental Industrial Hygienists (ACGIH) threshold limit value (TLV) and other applicable limits	________	________
6.	Physical and chemical characteristics, such as vapor pressure and flash point	________	________
7.	Physical hazards, including the potential for fire, explosion and reactivity	________	________
8.	Primary routes of entry into the body, such has inhalation, ingestion or skin absorption	________	________
9.	Acute and chronic health hazards, including signs and symptoms of exposure and medical conditions aggravated by exposure	________	________
10.	Carcinogenic hazard — National Toxicology Program (NTP) Annual Report on Carcinogens, International Agency for Research on Cancer (IARC) Monographs, or regulated by WISHA or OSHA **Note:** *If a chemical is not carcinogenic or if there is no information about its carcinogenicity, then information about item 10 doesn't have to be listed unless a blank is provided on the form.*	________	________
11.	Emergency and first aid procedures	________	________
12.	Precautions for safe handling and use including hygienic practices, repair and maintenance, protective measures, and spill/leak clean-up	________	________
13.	Exposure control measures such as engineering controls, work practices and personal protective equipment	________	________

Material Safety Data Sheet
May be used to comply with
OSHA's Hazard Communication Standard.
29 CFR 1910.1200. Standard must be
consulted for specific requirements.

U.S. Department of Labor
Occupational Safety and Health Administration
(Non-Mandatory Form)

IDENTITY *(As Used on Label and List)* **1**

Note: Blank spaces are not permitted. If any item is not applicable, or no information is available, the space must be marked to indicate that.

Section I

Manufacturer's Name **2**	Emergency Telephone Number
Address *(Number, Street, City, State, and ZIP Code)*	Telephone Number for Information
	Date Prepared **3**
	Signature of Preparer *(optional)*

Section II — Hazardous Ingredients/Identity Information

Hazardous Components (Specific Chemical Identity: Common Name(s))	OSHA PEL	ACGIH TLV	Other Limits Recommended	% (optional)
4	**5**			

Section III — Physical/Chemical Characteristics

Boiling Point		Specific Gravity (H_2O = 1)	
Vapor Pressure (mm Hg) **6**		Melting Point	
Vapor Density (AIR = 1)		Evaporation Rate (Butyl Acetate = 1)	

Solubility in Water

Appearance and Odor

Section IV — Fire and Explosion Hazard Data

Flash Point (Method Used) **6**	Flammable Limits	LEL	UEL

Extinguishing Media **7**

Special Fire Fighting Procedures

Unusual Fire and Explosion Hazards

(Reproduce locally)

Section V — Reactivity Data

Stability	Unstable		Conditions to Avoid
	Stable		

Incompatibility *(Materials to Avoid)* **7**

Hazardous Decomposition or Byproducts

Hazardous Polymerization	May Occur		Conditions to Avoid
	Will Not Occur		

Section VI — Health Hazard Data

Routes of Entry	Inhalation? **8**	Skin?	Ingestion?

Health Hazards *(Acute and Chronic)*

9

Carcinogenicity **10**	NTP?	IARC Monographs?	OSHA Regulated?

Signs and Symptoms of Exposure **9**

Medical Conditions
Generally Aggravated by Exposure

Emergency and First Aid Procedures **11**

Section VII — Precautions for Safe Handling and Use

Steps to Be Taken in Case Material is Released or Spilled

12

Waste Disposal Method

Precautions to Be Taken in Handling and Storing

Other Precautions

Section VIII — Control Measures

Respiratory Protection *(Specify Type)* **13**

Ventilation	Local Exhaust	Special
	Mechanical *(General)*	Other

Protective Gloves	Eye Protection

Other Protective Clothing or Equipment

Work/Hygienic Practices **12**

Math Review

Working with Feet and Inches

A foot is broken up into twelve equal parts called inches. On a tape measure, each inch is divided into sixteenths. To add or subtract, arrange the feet over feet, inches over inches, and any fractions above another. When carrying inches remember there are 12 inches in a foot, and to borrow a foot gives you 12 inches (not 10!).

6′ − 4 1/2″	8′ − 9″	
8′ − 6″	32′ − 11″	144′ − 0″
+14′ − 1/4″	+104′ − 6″	+ 2′ − 6″
28′ − 10 3/4″	144′ − 26″	146′ − 2″

To change feet into inches, multiply by 12. A board 8 feet long is 96 inches long (8 × 12 = 96). To convert 7′− 3″ into inches, multiply the 7 feet by 12 and add the 3 inches (7 × 12 = 84, 84 + 3 = 87 inches).

To change inches into feet, divide by 12. A board that is 60 inches long would be 5 feet (60 ÷ 12 = 5 feet). When converting inches to feet, any remainder is left as inches. A 75-inch board would be 6 feet 3 inches long (75 ÷ 12 = 6, remainder 3).

	31′ − 17″
13′ − 10″	32′ − 5″
− 6′ − 4″	−10′ − 9″
7′ − 6″	21′ − 8″

A yard is 3 feet long. There are 36 inches in a yard.

Practice Exercises

Convert to feet:

1. 144″
2. 21 yards
3. 120″
4. 96″
5. 84″
6. 48 yards

Convert to inches:

7. 12′
8. 8′ − 4″
9. 17′
10. 15′ − 7″
11. 4 yards
12. 4′ − 5″

Add:

13. 5′ − 6″
    +3′ − 4″

14. 3′ − 3″
    +7′ − 10″

15. 127′ − 11″
    52′ − 5″
    + 6′ − 10″

16. 6′ − 5″
    23′ − 10″
    +82′ − 9″

Subtract:

17. 24′ − 7″
    − 8′ − 5″

18. 12′ − 10″
    − 4′ − 8″

19. 8′ − 6″
    −4′ − 8″

20. 23′ − 6″
    − 7′ − 6″

Linear Measurements–Exercises

Linear measurements are simply how long something is—whether it is the length of a 2 × 4, the amount of trim boards needed, the perimeter of a lot, or the number of edge forms required.

1. A pile of lumber contains 198 boards 8 ft. long, 273 boards 10 ft. long, 135 boards 12 ft. long, and 18 boards 16 ft. long. How many feet of boards are in the pile?

2. How many feet of base is needed in a room that is 24 feet long and 18 feet wide (add all 4 walls)?

3. Oak is sold at $1.02 per foot. The job requires 8 pieces 10 ft. long and 17 pieces 8 ft. long. How many feet of oak are needed? How much will they cost?

4. The rabbits are getting into the garden. How much fencing is used to enclose a 24 ft. by 32 ft. area?

5. A carpenter can place 72 ft. of floor joists per hour. How many feet are done after 8 hours?

6. A roll of insulation is used for every three feet of framed wall. If there are 80 feet of wall to insulate, how many rolls must be **bought?**

7. Both edges of a wall top require chamfer. The pour is 64 feet long. If the chamfer is 8 feet long, how many are needed?

Fractions

A *fraction* is a part of a whole number. The fraction is made up of two numbers. The top number is called the *numerator* and the bottom number is called the *denominator.*

The denominator tells the number of equal parts the whole figure is being divided up into *(divisor).* The fraction 1/2 shows that the quantity will be divided into two equal parts.

The numerator tells how many of these parts are being used. The fraction 3/4 has the divisor of 4 indicating that the figure is divided into four equal parts. The numerator of 3 indicates that 3 of these parts are needed.

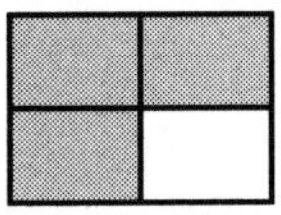

the shaded area = 3/4

A *proper* fraction has a numerator that is smaller than the denominator (3/4). An *improper* fraction has a larger numerator than the denominator (5/4). A *mixed* number has both a whole number and a fraction (2 1/4).

Measurements using fractions are easier if the fractions are written in the lowest terms. This does not change their value; for example, 1/2 = 2/4 = 4/8, etc. Try to think of it as if you were making change with money. Two quarters (50 cents) is equal to 5 dimes or 10 nickels or even 50 pennies. When reducing a fraction, find a number that will go into both the numerator and the denominator evenly. This number is called a *common factor.*

4/16 = 4/16 ÷ 4/4 = 1/4

The only number (besides 1) that can go evenly into both 4 and 16 is 4. When the numerator is one, the fraction cannot be reduced any further (4 is the common factor).

Adding Fractions

Adding fractions with the same denominator is basic addition. Add the numerators:

1/8 + 3/8 + 3/8 = **7/8** (1 + 3 + 3 = 7)

To reduce an improper fraction, divide the numerator (7) by the denominator (4). The answer is the whole number (1) and the remainder of 3 put into a fraction.

3/4 + 1/4 + 3/4 = 7/4 (3 + 1 + 3 = 7)

7/4 = 1 − 3/4 (7 ÷ 4 = 1 remainder 3)

```
  1
4)7
  4
  3
```

Lowest Common Denominator

When the denominators are not the same, you must first find the lowest common denominator (LCD). This is a number that can be evenly divided by each of the denominators you are adding or subtracting. Sometimes this can be done by just looking at the fractions. The LCD of 1/4, 1/2, and 1/8 would be 8 because it can be divided by 4, 2, and 8. If the LCD cannot be determined by inspection, then it may be found mathematically.

To find the LCD of the following fractions **3/4, 5/8, 11/32,** and **7/12,** there are two ways. Factoring is one way of determing the LCD; for example,

4 is 2 × 2,
8 is 2 × 2 × 2,
32 is 2 × 2 × 2 × 2 × 2, and
12 is 3 × 2 × 2.

The LCD would need to be 2 × 2 × 2 × 2 × 2 × 3 or **96.** All of the factors for each denominator must be found in the LCD; for example,

4 times 24 = **96,**
8 times 12 = **96,**
32 times 3 = **96,** and
12 times 8 = **96.**

To change the fractions to include the new denominator, remember to do the same thing to the numerator as you did to the denominator.

3 × 24 = 92	5 × 12 = 60	11 × 3 = 33	7 × 8 = 56
4 24 96	8 12 96	32 3 96	12 8 96

Another way of determining the LCD is through division.

Using the same fractions **3/4, 5/8, 11/32,** and **7/12:**

4, 8, 32, 12 Place the denominators in a line

2) 4, 8, 32, 12
2 4 16 6 Find the smallest number that can be divided into 2 or more of the denominators

Divide as many denominators by 2

2) 2 4 16 6
1 2 8 3 Repeat the division because 2 will go into 2 or more of the denominators

2) 1 2 8 3
1 1 4 3 Repeat again (if a number cannot be divided by the divisor, bring the number down) Now one is the only number that can be divided into the remaining numbers

2 × 2 × 2 × 1 × 1 × 4 × 3 = 96

Adding Mixed Numbers

When adding whole numbers, common fractions, and mixed fractions:

- add the fractions and reduce to lowest terms;
- add the whole numbers; then
- add the sums

Add 6, 5/8, and 1–3/4

6
5/8
+1 3/4 Write all numbers in a column

6
5/8 Reduce the fractions to LCD (8)
+1 6/8 Add the numerators and put
7 11/8 = 1 3/8 over the LCD
Reduce the fraction
Add the whole numbers

7
+1 3/8 Add the sums
8 3/8

When subtracting common fractions:

- reduce all fractions to their LCD; then
- subtract the numerators and put over the LCD and reduce to lowest terms.

Subtract 1/4 from 5/8:

1/4 = 2/8 Reduce fraction to LCD
5/8 – 2/8 Set up problem
5 – 2 = 3 Subtract the numerators
3/8 Put answer over LCD
5/8 – 2/8 = 3/8 This answer is in lowest terms

When subtracting fractions from whole numbers, you need to borrow 1 and change it into a fraction, much like changing a dollar bill into change for the phone.

- Subtract 1 from the whole number and change it into a fraction with the same denominator as the other fraction.
- Subtract the fractions, leaving the answer over the denominator.
- Reduce to lowest terms, if needed.
- Bring the remaining whole number down.

5
– 5/8 Subtract 5/8 from 5

4
5 8/8 Borrow 1 from 5 (4 plus 8/8 equals 5)
– 5/8 Change the 1 into a fraction

4
5 8/8
– 5/8
3/8 Subtract the fractions

4
5 8/8
– 5/8
4 3/8 Bring the whole number down

To subtract mixed numbers from whole numbers, just subtract the whole numbers after the fractions are done.

5	4 5 16/16	4 5 16/16
–2 3/16	–2 3/16	–2 3/16
	13/16	2 13/16

Ruler in Sixteenths

This is an example of an inch divided into sixteenths. The longest line divides the inch in half. The two next longest lines mark quarters. Four lines divide quarters into eights. The eight shortest lines separate the sixteenths. Practice reading a tape until you are comfortable reading 1/16, 1/8, 3/16, 1/4, 5/16, 3/8, 7/16, 1/2, 9/16, 5/8, 11/16, 3/4, 13/16, 7/8, and 15/16 without trying to count the lines. Start with the half-inch mark. The short mark before it is 7/16 and the mark right after it is 9/16.

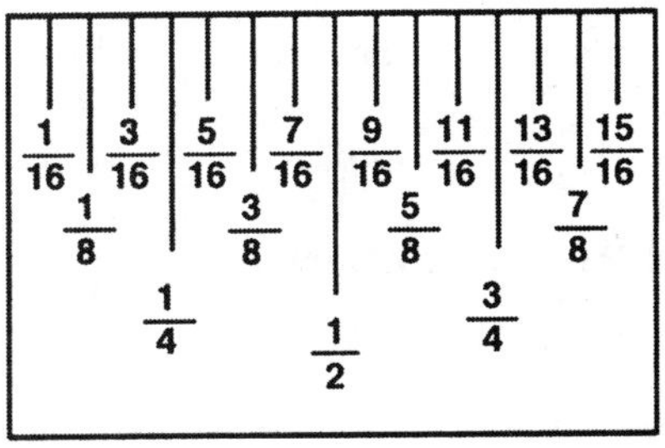

Try one-quarter — the mark before is 3/16 and the one after it is 5/16. The mark before three-quarters is 11/16 and the one after it is 13/16. The eighths seem to take care of themselves. Notice that the numerators (the top numbers) are all odd numbers. This keeps the fractions in their simplest form. Although 2/16 is a fraction of an inch, it is simpler to call it 1/8. The same is true of 8/16, it is equal to both 4/8 and 2/4, but we say 1/2. Eventually, a quick glance will tell you the measurement.

Practice Exercises with Fractions

Reduce to lowest terms:

1. $\frac{8}{12}$
2. $\frac{20}{32}$
3. $\frac{6}{8}$
4. $\frac{14}{16}$

Reduce to a mixed fraction:

5. $\frac{13}{8}$
6. $\frac{145}{12}$
7. $\frac{18}{4}$
8. $\frac{96}{8}$

Change to sixteenths:

9. 1/4
10. 3/8
11. 1/2
12. 7/8

Add the fractions and reduce all answers to lowest terms:

13. $\frac{3}{8} + \frac{2}{8} + \frac{1}{8} =$
14. $\frac{5}{16} + \frac{3}{16} + \frac{7}{16} =$
15. 1/2
 1/4
 +3/4
16. 3/8
 1/4
 +5/16
17. 7/8
 5/16
 +13/32
18. 1/8
 3/16
 +5/12
19. 15 1/4
 32 3/4
 +18 1/4
20. 23 7/8
 13 1/2
 +52 5/16
21. 2 3/8
 5/16
 +3 5/64

Subtracting Fractions

Solve the problems and reduce fractions to lowest terms:

1. 5/8
 −3/8
2. 9/16
 −3/16
3. 3/4
 −1/4
4. 21/32
 −17/32
5. 5/8
 −3/16
6. 15/16
 − 3/4
7. 3/4
 −4/32
8. 31/32
 − 5/8
9. 2
 − 3/16
10. 4
 − 5/8
11. 7
 − 1/4
12. 3
 − 11/16
13. 11 7/16
 − 8 5/16
14. 22 5/8
 −10 3/8

15. 2 3/8
 −1 1/8

16. 12 3/16
 −8 5/8

17. 18
 − 4 7/8

18. 11 1/2
 − 2 3/4

19. 68 1/2
 −57 9/16

20. 47 1/4
 −19 9/32

21. 7 1/16
 −6 1/2

Multiplying Fractions

$\frac{1}{4} \times \frac{3}{8}$ can be worked as $\frac{1 \times 3 = 3}{4 \times 8 = 32}$

Remember that when you multiply fractions, the answer gets smaller. Dividing fractions produces a larger answer.

The word *of* is sometimes used in place of the multiplication sign (×).

$$\frac{1}{2} \text{ OF } \frac{3}{4} = \frac{3}{8}$$

When multiplying mixed numbers, first change them into improper fractions.

$$4\frac{1}{2} \times 2\frac{3}{4} \text{ becomes } \frac{9}{2} \times \frac{11}{4} = \frac{99}{8}$$

(4 × 2 + 1) there are 9 halves in 4 1/2
(2 × 4 + 3) there are 11 quarters in 2 3/4

After multiplying, reduce the answer back to a mixed number

$99 = 12\frac{3}{8}$ (99 ÷ 8 = 12 with a remainder of 3)

```
   12
 8)99
   8
   19
   16
    3
```

When using cancellation, you are dividing the equation by 1 (2/2, 4/4, etc.).

Multiply 16, 3/8, and 1/2:

$$\frac{16}{1} \times \frac{3}{8} \times \frac{1}{2}$$

```
 1
 2
16     3     1
 —  ×  —  ×  —
 1     8     2
       1     1
```

8 will divide into 16 and 8
2 will divide into 2 and 2

$$\frac{1}{1} \times \frac{3}{1} \times \frac{1}{1} = 3 \text{ answer}$$

Dividing Fractions

To divide fractions you must:

- invert the divisor (flip the second fraction upside-down);
- multiply across the top (numerators) and the bottom (the denominators); and
- reduce to lowest terms.

1/2 divided by 1/4 (this is asking how many times 1/4 will go into 1/2):

1/2 ÷ 1/4 =

1/2 × 4/1 — Invert the divisor

$\frac{1}{2} \times \frac{4}{1} = \frac{4}{2}$ — Multiply the fractions

4/2 = 2 — Reduce into lowest terms

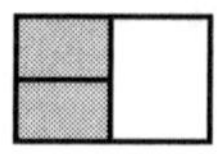

1/4 will go into 1/2 2 times

When dividing mixed numbers, change them both to improper fractions, then invert the divisor and multiply as above:

4 1/2 ÷ 2 3/4 =
9/2 ÷ 11/4 =

```
       2
9/2 × 4/11 = 18/11 = 1 7/11
  1
```

Practice Exercises

Multiplying and dividing fractions — reduce to lowest terms:

1. 1/2 × 3/4 = ____________
2. 1/4 of 1/4 = ____________
3. 3/8 × 5/16 = ____________
4. 3/5 × 20 = ____________
5. 3/8 × 1/16 = ____________

6. 5/8 of 3/5 = __________

7. 3 1/2 × 1 1/16 × 1/2 = __________

8. $\frac{1}{8} \times \frac{5}{16} \times \frac{4}{7} \times \frac{3}{5} =$ __________

9. 3 1/4 × 4 5/8 × 2 1/2 = __________

10. 11 1/4 × 2 1/2 × 6 = __________

11. 3/8 ÷ 1/2 = __________

12. 7/16 ÷ 7/8 = __________

13. 4 1/3 ÷ 6 2/3 = __________

14. 16 ÷ 1 5/8 = __________

15. 2 1/8 ÷ 3/4 = __________

16. 5 ÷ 5/8 = __________

Word Problems with Fractions

Show your work (or the equations used) with a calculator.

1. A carpenter makes $963.60 for a 40-hour week. What is his hourly wage?

2. One thousand one hundred and forty-four sheets of drywall are used to hang an office building. If each office averages 52 sheets, how many offices are there?

3. If it takes 7/8 yard of concrete to pour a column, how much will it take to pour 24 columns?

4. Determine the number of hours required to lay 847 square feet of subflooring at the rate of 82 feet per hour.

5. The interior walls of a house are framed 3 1/2 inches. Both sides are covered with 1/2-inch drywall. How thick are the partitions?

6. The exterior walls of a house are 5 1/2 inches thick. Insulation board 1/2 inch thick is installed on the outside. This is covered with 5/8-inch cedar. After 1/2-inch drywall is hung inside, what is the total thickness of the wall?

7. How many pieces of 3/4-inch plywood are in a stack that is 89 1/4 inches tall?

8. The cost of a new garage is $14,824. What is the total cost to build 12 of these garages?

9. There are 806 fourpenny finish nails in a pound. How many are there in 19 1/4 pounds?

10. If there are 670 shingles per square (100 square feet) when laid 5 inches to the weather, how many shingles are needed to cover 32 squares?

11. A rough piece of oak is 2 1/16 inches thick. The planer takes off 5/32 inch from each side. How thick is it now?

12. A carpenter can frame 1,000 board feet of wall in 21.4 hours. How many feet of wall can be done after a 40-hour work week?

13. If the carpenter used 13 pounds of nails during the week and there are 332 nails per pound, how many nails were driven?

14. Each column takes 1 1/3 sheet of 3/4-inch plywood. How much plywood is needed for 24 columns?

15. A stack of lumber has 948 2 × 4s. If 1/4 of the stack is used, how many are gone?

16. If 2/3 of the remaining lumber from problem 15 is used, how many are left?

17. The scale for plans of a building is 1/4-inch = 1 foot. How long is the building if it scales 23 3/4 inches?

18. A carpenter must divide a layout stick 75 inches long into 12 equal spaces. How long is each space?

Board Feet

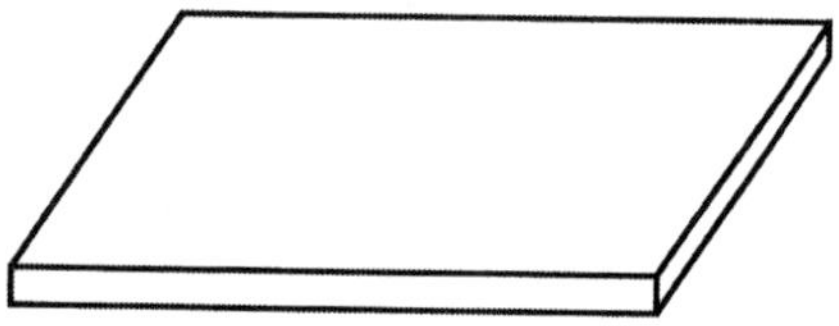

A *board foot* is 1″ thick by 12″ wide and 1′ long. Lumber is often sold at a price per thousand board feet. **M** is the Roman numeral for 1000.

$$\text{Board feet} = \frac{\text{Thickness (in inches)} \times \text{Width (in inches)} \times \text{Length (in feet)}}{12}$$

Example: To figure the number of board feet in a 1″ × 8″ × 16′:

$$\frac{1 \times 8 \times \overset{4}{16}}{\underset{3}{12}} = \frac{32}{2} = 10\ 2/3 \text{ board feet}$$

If more than one of the same sized boards are being figured, multiply it.

$$\frac{\text{Number of Pieces} \times T \times W \times L}{12}$$

Example: To figure the number of board feet in 24 pieces of 2″ × 4″ × 16′:

$$\frac{\overset{2}{24} \times 2 \times 4 \times 16}{\underset{1}{12}} = 256$$

If lengths are given in inches, multiply the denominator by 12.

Example: To figure the number of board feet in 4 pieces of 1″ × 6″ × 18″:

$$\frac{\overset{1}{4} \times 1 \times \overset{1}{6} \times \overset{3}{18}}{\underset{\underset{1}{2}}{12} \times \underset{\underset{1}{2}}{12}} = 3 \text{ board feet}$$

Board Feet Exercises

NOTE: Lumber less than 1″ thick is counted as a full inch. Greater than 1″ thick, standard thicknesses are 1 1/4″, 1 1/2″, 2″, 3″, 4″, 5″, 6″, etc.

For *odd widths,* count the next standard width larger. Standard widths are 2″, 3″, 4″, 5″, 6″, 8″, 10″, 12″, etc.

Find the number of board feet in the following:

1. 15 pieces of 1/2″ × 4″ × 16′ __________
2. 54 pieces of 3/4″ × 9 1/4″ × 14′ __________
3. 84 pieces of 5/8″ × 8″ × 16′ __________
4. 47 pieces of 3/4″ × 3 1/2″ × 18′ __________
5. 17 boards, 1″ × 12″ × 10′ __________
6. 9 boards 1 1/2″ × 22″ × 16′ __________
7. 7 planks 2″ × 12″ × 14′ __________
8. 10 pieces 2″ × 6″ × 12′ __________
9. 2 pieces 1″ × 2″ × 22″ __________
10. 7 pieces 1″ × 2 1/4″ × 17″ __________
11. 25 pieces 3/4″ × 3″ × 12′ __________
12. 60 linear feet of 1″ × 4″ __________
13. 260 linear feet of 2″ × 6″ __________
14. 140 linear feet of 2″ × 4″ __________
15. 68 linear feet of 1″ × 3″ __________

Area

Area of a square is side times side or side squared. The answer is left as square inches (sq. in.) or square feet (sq. ft.). It is **never** changed to feet and inches, as this would be a length, not an area.

Find the area of this 3-inch square.

$$3 \times 3 = 9$$

(3 inches times 3 inches equals 9 square inches)

When finding the area of a square with sides expressed in feet and inches, either change the length into inches and multiply to get an answer in square inches, or convert the inches into decimals of a foot and multiply to get an answer in square feet.

AREA of a square with sides 2′ 6″:
$2 \times 12 = 24 \quad 24 + 6 = 30 \quad 30'' \times 30'' = 900$ sq. in.

AREA of a square with sides 2′ 6″:

$6 \div 12 = .5 \quad 2 + .5 = 2.5 \quad 2.5' \times 2.5' = 6.25$ sq. ft.

A square foot has 144 square inches ($12'' \times 12'' = 144''$).

To change an answer from square inches into square feet, divide by 144.

900 sq. in. $\div$ 144 = 6.25 sq. ft.

To change an answer from square feet into square inches, multiply by 144.

6.25 sq. ft. $\times$ 144 = 900 sq. in.

It is usually easier to work with smaller numbers, so an answer in square feet would be preferred.

Area of Plane Figures

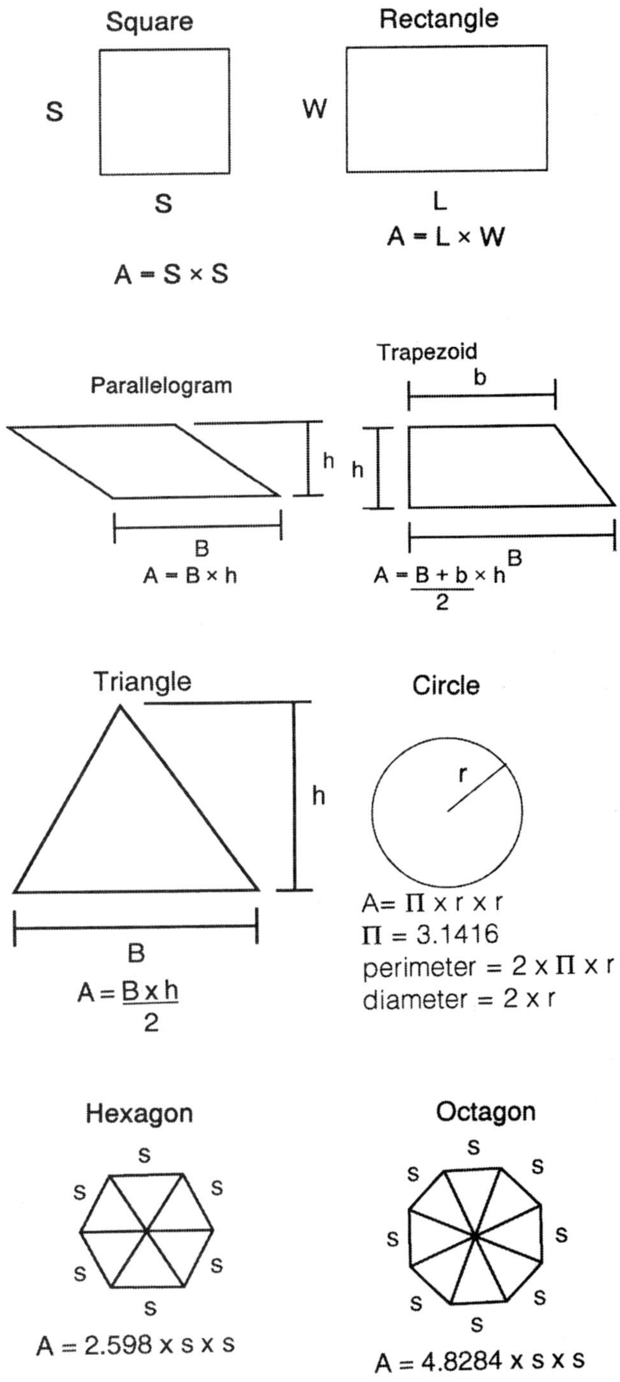

Area Exercises

Find the area of the following squares:

1. Side = 12′ 0″ __________
2. Side = 15′ 0″ __________
3. Side = 28″ __________

4. Side = 42″ __________

5. Side = 1′ 9″ __________

6. Side = 4′ 5″ __________

7. Side = 24′ 6″ __________

8. Side = 36′ 3″ __________

9. Side = 5′ 6″ __________

10. Side = 3′ 9″ __________

Find the area of the following rectangles:

11. Width = 7′ Length = 5′ __________

12. Width = 8′ Length = 11′ __________

13. Width = 8″ Length = 17″ __________

14. Width = 19″ Length = 47″ __________

15. Width = 3′ 1″ Length = 4′ 7″ __________

16. Width = 4′ 10″ Length = 7′ 2″ __________

17. Width = 27′ 6″ Length = 37′ 6″ __________

18. Width = 14′ 3″ Length = 21′ 3″ __________

19. Width = 11′ 9″ Length = 21′ __________

20. Width = 15′ 0″ Length = 18′ 6″ __________

Find the areas of the following circles:

21. Radius = 6′ 0″ __________

22. Diameter = 20″ __________

23. Radius = 9″ __________

24. Radius = 14′ 0″ __________

25. Radius = 3′ 6″ __________

26. Diameter = 3′ 0″ __________

27. Diameter = 8′ 0″ __________

28. Radius = 11″ __________

29. Radius = 18″ __________

Find the area of the following triangles:

A = Altitude B = Base

30. A = 8′ B = 15′ __________

31. A = 7″ B = 13″ __________

32. A = 17″ B = 22″ __________

33. A = 2′ 5″ B = 4′ 1″ __________

34. A = 4′ 10″ B = 4′ 5″ __________

35. A = 10′ 6″ B = 15′ 4″ __________

Find the area for the following parallelograms:

B = Base h = height

36. B = 20″ h = 10″ __________

37. B = 4′ 3″ h = 2′ 4″ __________

38. B = 13′ 0″ h = 11′ 2″ __________

39. B = 22″ h = 13″ __________

40. B = 3′ 4″ h = 5′ 5″ __________

41. B = 5′ 2″ h = 4′ 10″ __________

Find the area of the following trapezoids:

42. B = 8″ b = 6″ h = 3″ __________

43. B = 12′ b = 8′ h = 4′ __________

44. B = 8′ 10″ b = 4′ 3″ h = 3′ 2″ __________

45. B = 7′ 4″ b = 5′ h = 2′ 11″ __________

46. B = 18′ b = 13′ 3″ h = 10′ __________

47. B = 44″ b = 32″ h = 19″ __________

Find the area of the following hexagons:

48. Side = 4″ __________

49. Side = 6′ __________

50. Side = 3′ 4″ __________

51. Side = 7′ 11″ __________

Find the area of the following octagons:

52. Side = 19″ __________

53. Side = 3′ 7″ __________

54. Side = 28′ ____________

55. Side = 9′ 11″ ____________

Volume

The volume of a cube is found by multiplying side times side times side, or side cubed.

Volume is area times height or thickness:

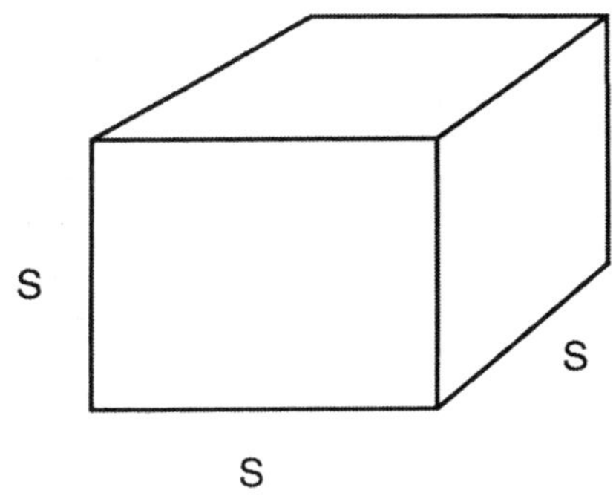

Volume is always expressed in cubic inches, cubic feet, or cubic yards. There are 1728 cubic inches in a cubic foot (12 × 12 × 12 = 1728).

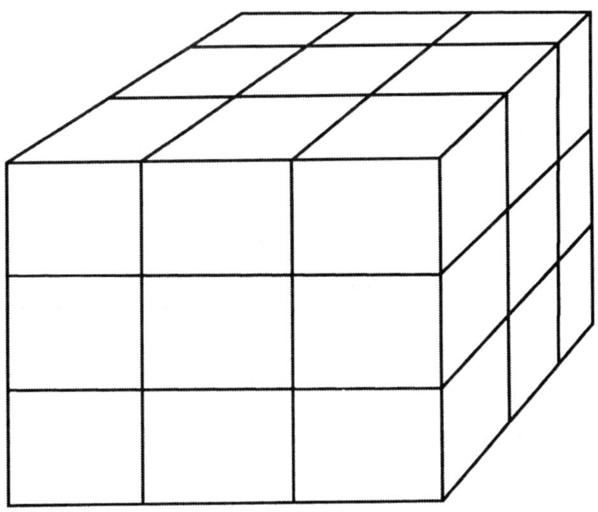

There are 27 cubic feet in a cubic yard (3 × 3 × 3 = 27).

Concrete is ordered by the cubic yard. It is easier to use cubic feet than inches when figuring concrete volume. There are 46,656 cubic inches in a cubic yard.

To determine how many yards of concrete are in a form, divide the volume in cubic feet by 27 (the number of cubic feet in a cubic yard).

Volume of Solid Figures

Rectangular Solid

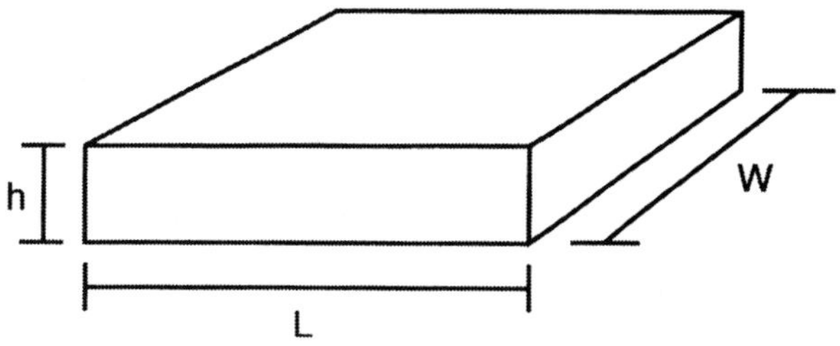

$$V = L \times W \times h$$

Cylinder

$$V = \Pi\, r \times h$$

Cone

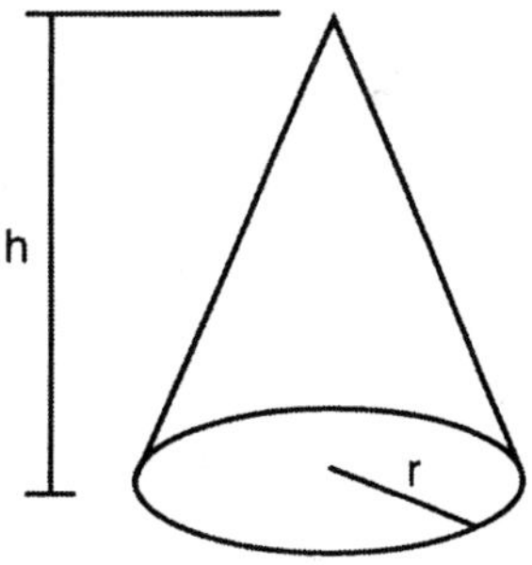

$$V = \frac{\pi \times r \times r \times h}{3}$$

Pyramid

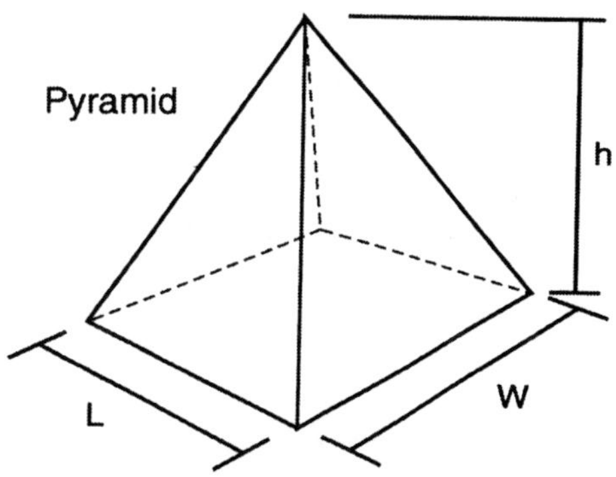

$$V = \text{area of base} \times \frac{h}{3}$$

Frustum of a Cone

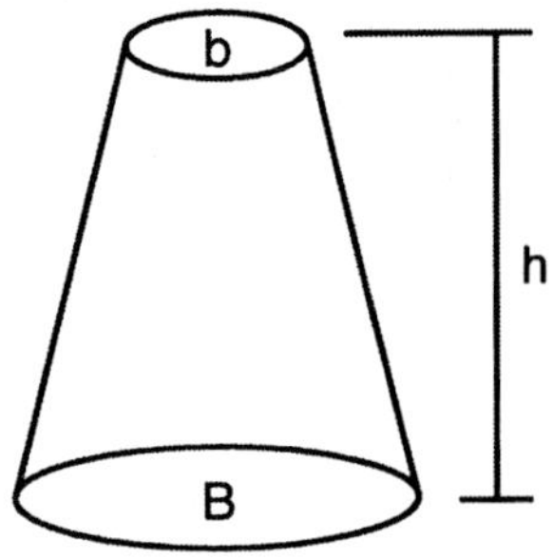

Frustum of a Pyramid

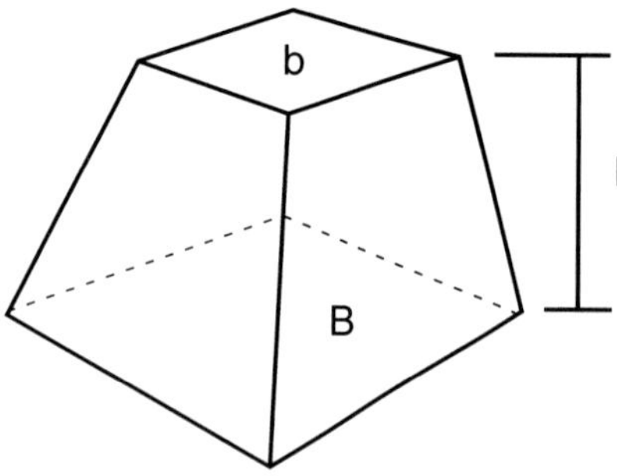

$$V = (B + b + \sqrt{B \times b}) \times \frac{h}{3}$$

Where B = area of lower base
b = area of upper base

Volume Word Problems

Solve the following word problems:

The following six questions are based on Figure 1, which represents a basement, and Section A, which is a cross section of the footing, foundation wall, and slab. Allow five feet on all sides when excavating to provide for forming.

1. How many cubic yards of dirt were removed, assuming 3 feet of extra dirt is also removed to a depth of 8 feet?

2. How many cubic yards of concrete are there in the footing?

3. How many cubic yards of concrete are there in the foundation walls?

4. How many cubic yards of concrete are there in the slab?

5. What is the cost of the concrete in #2, 3, and 4 at $75.00 per cubic yard?

6. How many dump truck loads of earth are removed for excavation? Each truck can carry 9 cubic yards.

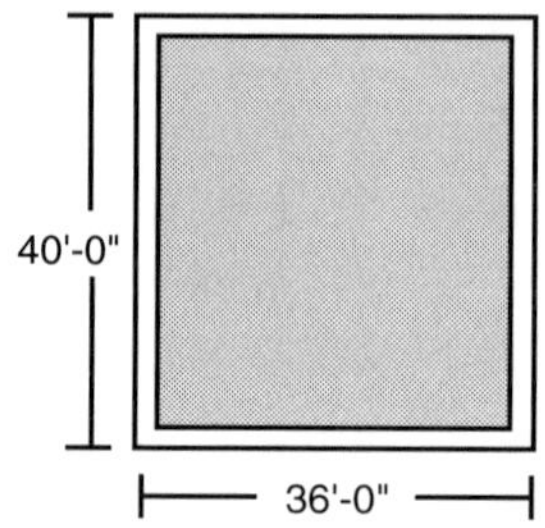

Figure 1

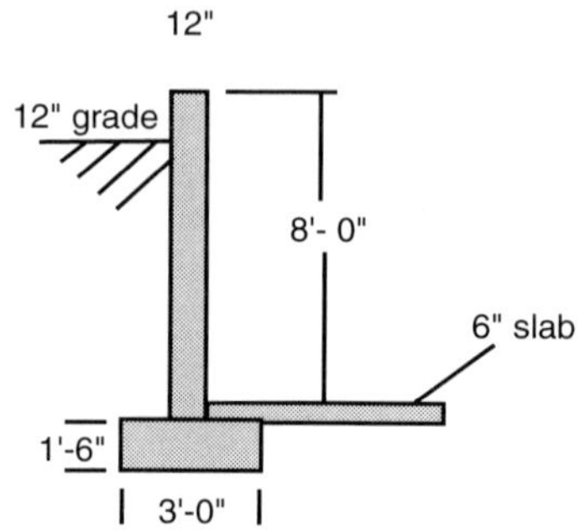

Section A

Solve the following word problems:

7. How many linear feet of edge forms are needed to form the walk in Figure 2?

8. How many cubic yards of concrete are needed to pour the walk 6″ deep?

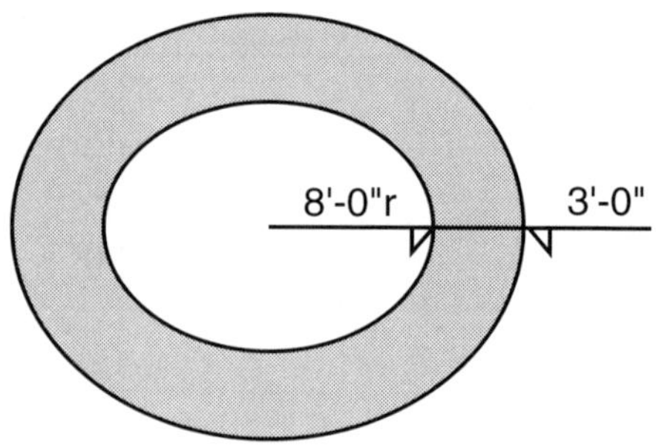

Figure 2

9. An accent wall extends to the ceiling height of 16 feet. It has the shape of a trapezoid. It is 55 feet long at the floor and 40 feet long at the ceiling (see Figure 3).

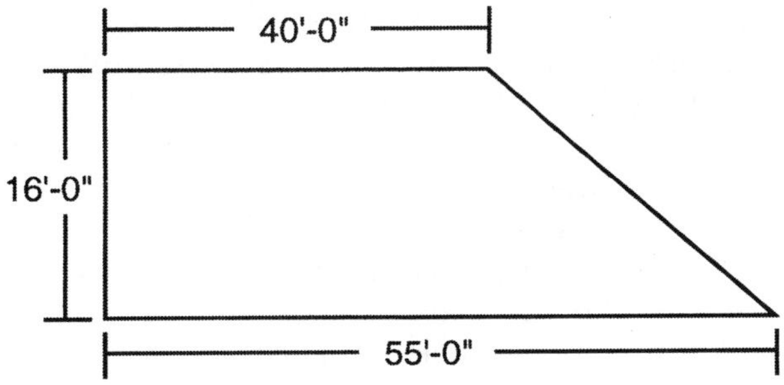

Figure 3

a. How many square feet does the wall contain?

b. How many sheets of 4 × 12 drywall are needed to hang the wall if 10% was added for waste?

c. How many sheets of 4 × 8 drywall would be needed?

Solve the following word problems:

10. What is the volume of the column and capital in Figure 4?

11. How many yards of concrete would it take to pour 10 pier footings in Figure 5?

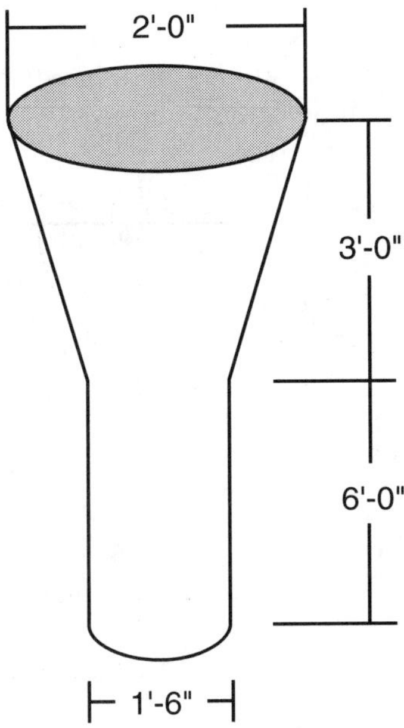

Figure 4

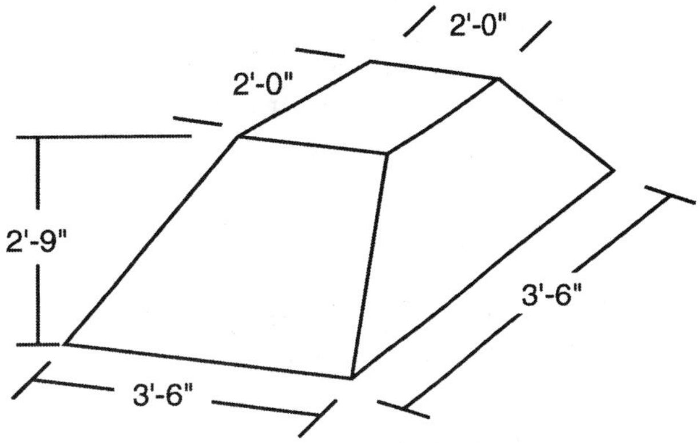

Figure 5

12. Find the cost of 16 cubic yards of gravel, weighing 106 pounds per cubic foot, at $9.95 per ton.

13. How many tons of gravel, weighing 106 pounds per cubic foot, will a truck carry if the inside body measurements are 5′ 9″ wide, 14′ 6″, long, and 3′ 0″ tall (load will be leveled to the top of the sides)?

Triangles

All shapes with three or more straight sides are made up of triangles. When a vertical line meets a horizontal line, a right angle is formed. A right angle is 90 degrees. The right triangle is the most common angle used by carpenters. The longest side is called the *hypotenuse*.

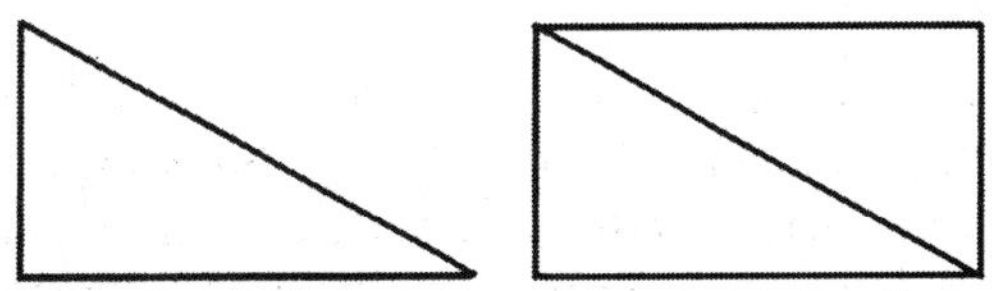

The right triangle is one half of a rectangle or square.

Rectangles have two sides that are equal in length and the two ends are equal to each other. All four inside corners are 90 degrees. A square is a rectangle with four equal sides.

3–4–5

Carpenters check for square either by pulling a 3–4–5 to determine a right angle or by measuring across the diagonal of a building. When using a 3–4–5 to determine a right angle, the carpenter is applying the *Pythagorean Theorem*. Pythagoras was a Greek philosopher and mathematician who lived about 2,500 years ago. His theorem (a math rule of relations) states that in a right angle, the square of the hypotenuse is equal to the sum of the squares of the other two sides. This is written as a formula: $c^2 = a^2 + b^2$

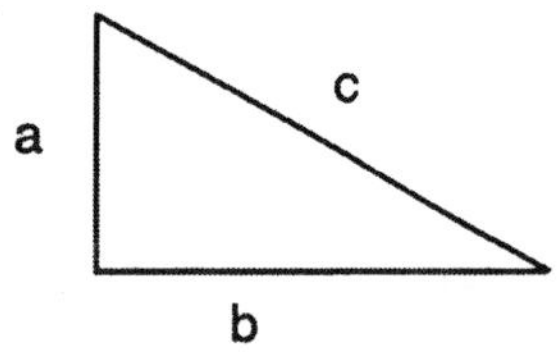

where **c** is the length of hypotenuse, and **a** and **b** are the lengths of the other two sides. If the two sides are known, the missing length can be found by substituting the values in the formula.

5 × 5 = 3 × 3 + 4 × 4 25 = 9 + 16

Over 4,000 years ago, the Egyptians discovered the 3–4–5 triangle. A length of rope was knotted into 12 equal spaces. Three stakes were then driven at the corners spaced 3, 4, and 5 knots apart.

To increase the size of the triangle, you must multiply all parts of the triangle by the same number. For example 6–8–10 would be the result of multiplying by 2. For even greater distances, 30–40–50 could be used.

The sum of all three angles of any triangle equals 180 degrees. If the two angles are known, the other can be found by subtraction.

For example, a right triangle has a 60-degree angle:

180 − (90 + 60) =
180 − 150 = 30 (the missing angle is 30 degrees.)

The total (180) *minus the two known angles* (90 + 30) *equals the other angle.*

A right triangle with two 45-degree angles is used frequently by carpenters. Because the angles are the same, the length of the two sides are also equal to each other. To brace a 2-foot wall with 45-degree kickers, measure 2 feet from the wall and drive the stake. The brace will be at a 45-degree angle at the top of the wall and at the stake. A quick reference to determine the length is found on the back tongue of your framing square.

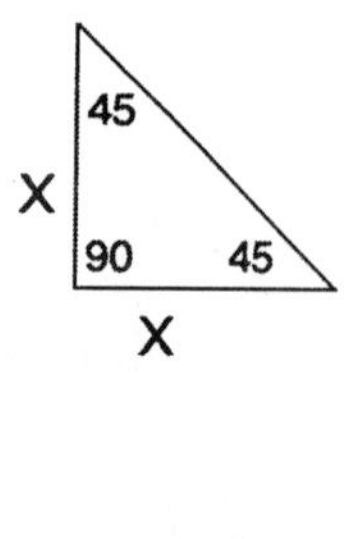

Using a calculator with a square root key, you can determine the hypotenuse of any size right triangle. If one side is 7 feet and the other side is 9 feet, the formula would be: **7 × 7 + 9 × 9 = 49 + 81 = 130.** Now hit the square root key = **11.401756.** This can be rounded off to 11.40, or rounded up to 11.402. The hypotenuse changes less than 1/16″. The square root can also be computed by longhand math. However, it is not necessary for students to use this outdated practice.

In the field this can be done by measuring the diagonals of a rectangle (remember that these are the hypotenuse of two right angles). Adjust the corners in or out until the diagonals are equal. Subtract the smaller diagonal from the larger one and divide the difference in half to find out how far out of square it is (how far to move the corner).

Definitions

A straight angle equals 180° or a straight line.

Intersecting lines form opposite equal angles (angle 1 = angle 2).

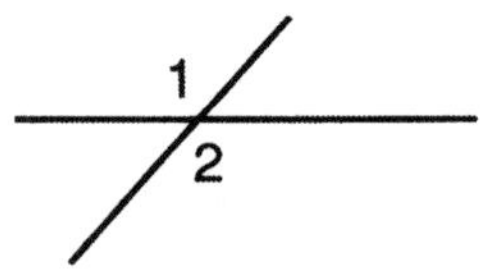

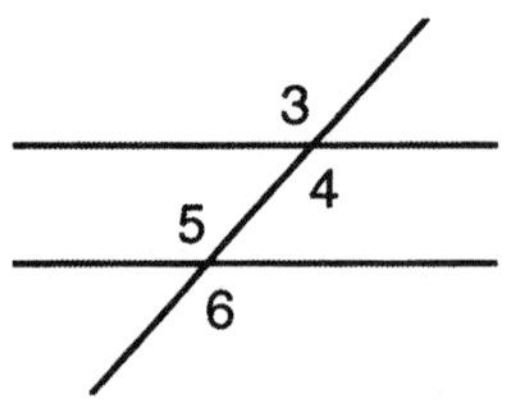

Parallel lines (lines which have a constant distance separating them) cut by an intersecting line form alternate equal angles, so that angle 3 = angle 4 = angle 5 = angle 6.

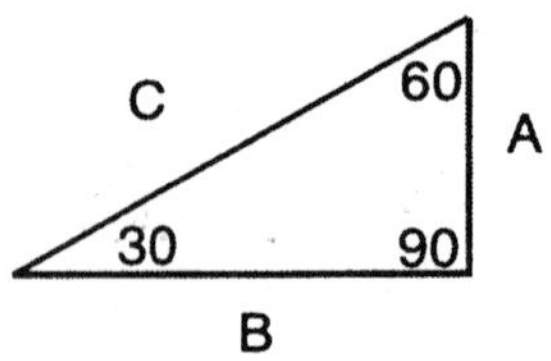

The side opposite a 30° angle of a 30–60 degree triangle will equal 1/2 of the hypotenuse.
A = 1/2 C.

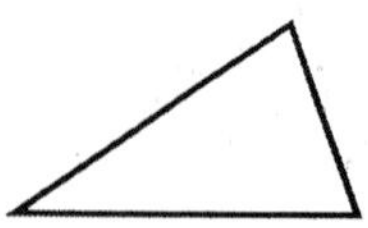

An acute triangle has all angles measuring less than 90 degrees.

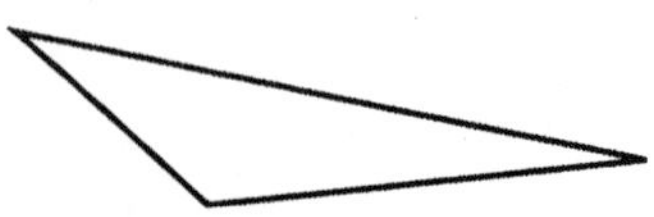

An obtuse triangle has an angle that is greater than 90 degrees.

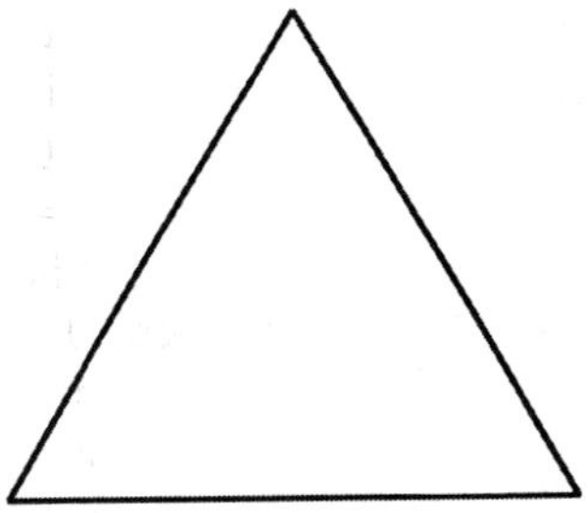

An equilateral triangle has three equal sides and three equal angles of 60 degrees each.

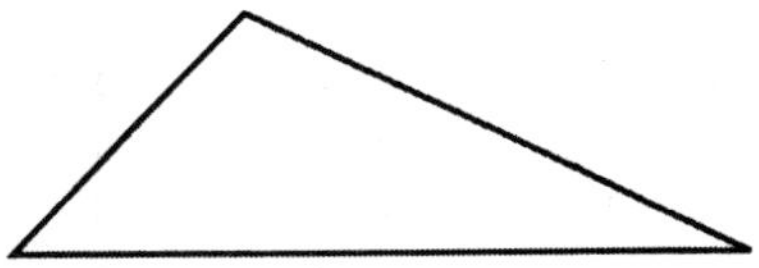

A scalene triangle has sides of different lengths and three different angles.

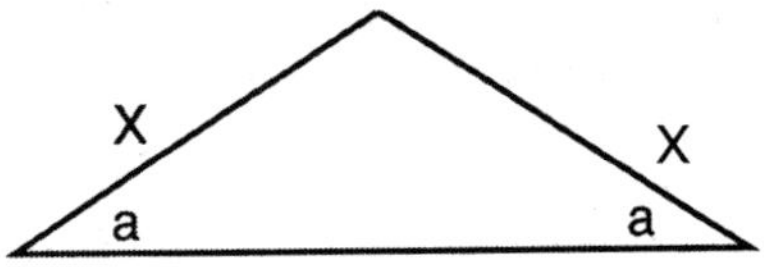

An isosceles triangle has two equal sides and two equal angles.

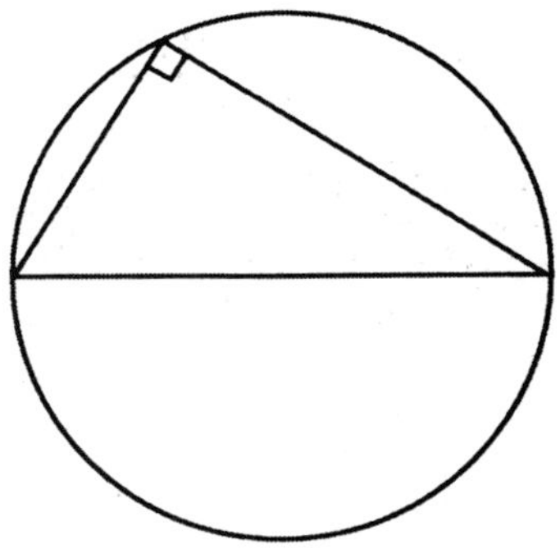

An angle inscribed in a semicircle will always be a right angle.

Circles and Angles

In layout work carpenters use angles of various sizes. To work intelligently, one must have knowledge of what an angle is and how to measure it. The measure of an angle is based on the opening between two lines coming out from a center point called a *vertex*. The size of the angle is this opening, not the length of the lines.

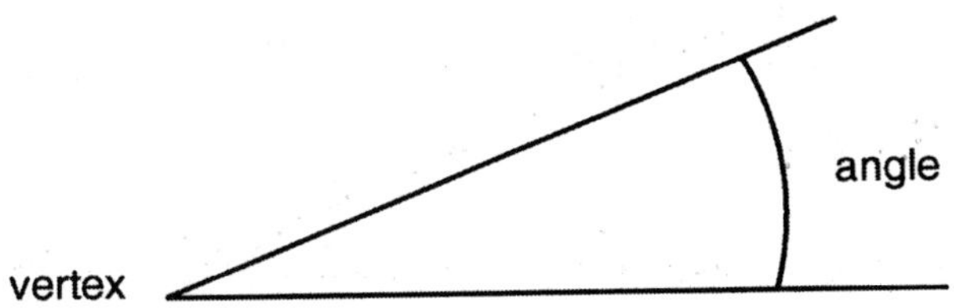

Degrees, Minutes, and Seconds

Angles are measured in terms of degrees, minutes, and seconds. A degree is found by dividing a circle into 360 equal parts. The symbol used to identify a degree is a small circle placed slightly above and to the right of the number; so then 25° means 25 degrees.

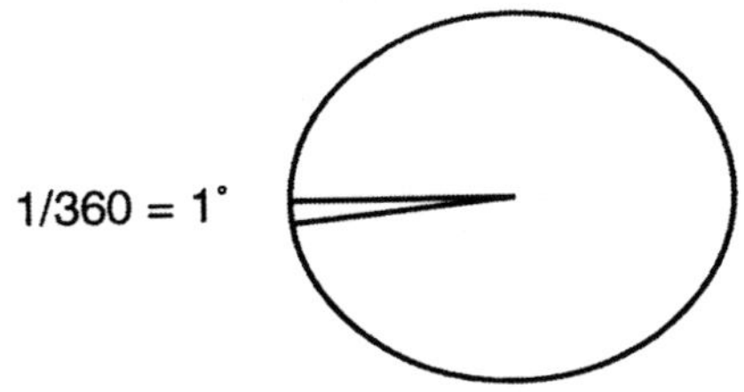

A degree is divided into 60 equal parts called *minutes*. Minutes are symbolized by ′. Just like a clock, a minute has 60 seconds. Seconds are shown with a ″ mark. This measurement, 48°-32′-56″ would be read forty-eight degrees, thirty-two minutes and fifty-six seconds.

Angles can be added and subtracted. To add 3°-35′-29″ with 12°-6′-20″, line the degrees up with the degrees, the minutes with the minutes, and the seconds with the seconds. Add each column separately.

```
  3°-35′-29″
+12°-6′-20″
-----------
 15°-41′-49″
```

NOTE: When adding minutes and seconds that total more than 60, subtract multiples of 60 and carry the multiples to the higher value.

Example:

```
 117 seconds          144 minutes
-  60 (1 × 60)       -120 (2 × 60)
--------------       -------------
  57 remainder         24 remainder
```

117 seconds = 1 min. 57 sec.
144 minutes = 2 degrees 24 min.

To subtract 12°-10′-14″ from 30°-27′-38″, again, line up the degrees, minutes, and seconds.

```
 30°-27′-38″
-12°-10′-14″
------------
 18°-17′-24″
```

NOTE: When you borrow during subtraction, remember that 1 degree = 60 minutes and 1 minute = 60 seconds.

Practice Exercises with Angles

Find the missing angle of the triangle.

1. 20, 40 __________

2. 80, 40 __________

3. 90, 60 ____________

4. 30, 60 ____________

5. 70, 50 ____________

6. 58, 79 ____________

7. 65, 90 ____________

8. 90, 50 ____________

Multiples of 3–4–5 Triangles

9. 6 – 8 – ____

10. 9 – ____ – 15

11. ____ – 40 – 50

12. 12 – 16 – ____

13. 36 – ____ – 60

14. ____ – 28 – 35

15. 24 – 32 – ____

16. 15 – ____ – 25

Solve:

17. 36°-50′-40″
+ 167°-9′ -40″

18. 91°-47′-30″
+ 27°-39′-42″

19. 107°-18′-20″
+ 77°-27′-15″

20. 180°
– 88°-35′-7″

21. 300 minutes = ______ degrees

22. 270 seconds = ______ minutes ______ seconds

23. 184 minutes = ______ degrees ______ minutes

Assume all spaces are equal:
Determine the degrees of the angles in each circle
A = ______ degrees

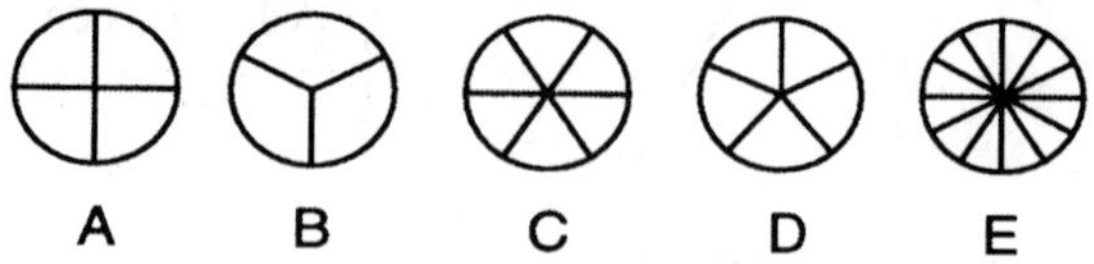

B = ______ degrees

C = ______ degrees

D = ______ degrees

E = ______ degrees

Solve for the missing angles:

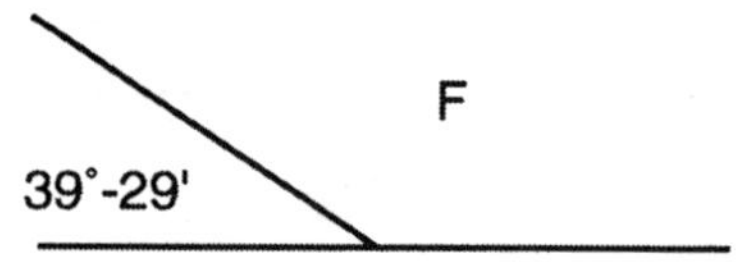

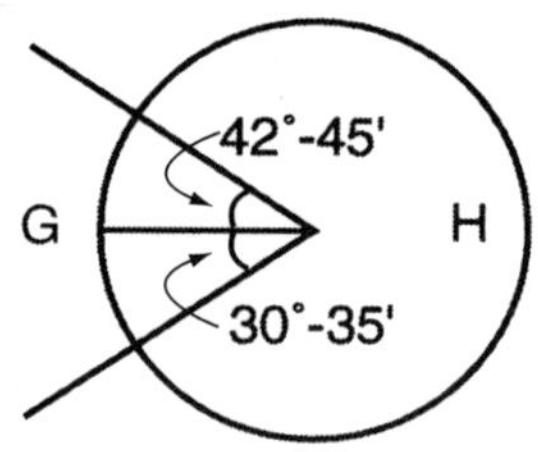

F = ______ degrees ______ minutes

G = ______ degrees ______ minutes

H = ______ degrees ______ minutes

Find the values of the missing angles:

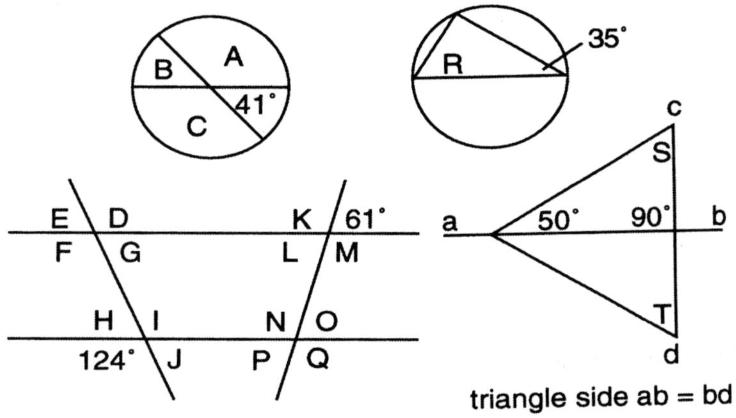

A ____________

B ____________

C ____________

D ____________

E ____________

F ____________

G ____________

H ____________

I ____________

J ____________

K ____________

L ____________

M ____________

N ____________

O ____________

P ____________

Q ____________

R ____________

S ____________

T ____________

Blueprints

When introducing blueprints to your students, use the general-to-specific approach. This is quite similar to focusing in a camea lens. After reading Unit 8 (Chapters 20–23) use this classroom exercise to help the students understand how to *focus* in on details of prints. Have copies of a full set of prints available; students can share, but each must have access to clearly see every sheet.

Plans of a recent remodel or addition in the area will help the students visualize the structure from the prints. Photos or a field trip to see the actual structure will also enhance their understanding. Slides of the completed project can be used for years.

Start with the plot plan. Discuss basics such as the nearest cross streets, the address, north orientation, the scale, location and name of main roads, the set back, and interesting features of the area. Ask the students questions and help them locate the answers on the plans. Depending on the amount of feedback, this session could be relatively brief.

Move on to the foundation plan (if included). Ask questions about footing length, wall thickness, excavation, reinforcing steel; also note hidden lines of the footings and any elevations or bench mark given.

Continue through all pages of the plans until sectionals, and ultimately details, are covered. By the time a student can understand a detail, where it came from, and what page to find it on, they have *zoomed in.* Zooming out (starting with a detail) does not work as well because students must wait to get the big picture.

Planning and Scheduling

This is another classroom exercise that gets the students involved in group discussion. Students do not need to have any field experience to be successful. This has been done with people who have neither been on a job site nor have read prints. The instructor can make up any conditions that they want; for example, the financing may be preapproved, there may be a 90-day time limit on construction, all environmental impact studies are complete, the permit is in hand, or anything else that might come up.

Start the discussion with the statement, "Today, we will build a house." Begin the planning process. Have the class *brainstorm* as many activities involved with building a house. Write these on a board or overhead so everyone can see. During the brainstorming, the ideas will be presented in random order. The instructor may give direction as needed by asking questions. For example, if backfilling has been overlooked, you might ask, "How are you going to protect from falls into the trenches?" Break down the tasks as needed for proper scheduling.

Now assign numbers to the tasks, with number one being the first item to be done. Continue through the list. At this point some items overlooked may become apparent — add them. The students should conclude that some tasks can be done simultaneously.

Agree as a group how much time each task will take. The background of the students will determine how much input the instructor will have. Keep the class focused. Write the allotted time behind the task.

You are now ready to begin scheduling these activities. Arrange the tasks in the order that has been assigned to them. An additional board may be needed. Across the top write out the project's time line (allow a 5-day work week, showing all work days). Fill in a basic bar graph. Show where multiple tasks are being done by subs at the same time. Talk about the coordination needed to get the job done and the interdependency of the crafts. The job goes smoother when everyone works together.

Include a discussion about material delivery and on-site storage. Long lead or special order items should be covered, since they can easily upset the schedule if they are late. The influence weather and the time of the year have on the timeliness of completion should be covered. Do not forget Murphy's Law — share some problems you have encountered in the field and how they were solved.

Try to get samples of schedules from job sites. They can be as basic as a one-week bar graph, more involved such as an Arrow Diagram, or as complicated as a Critical Path (CPM). The more examples made available to the students, the better. The advisory committee may be able to assist you in getting copies. Have a guest speaker teach planning and scheduling, if available. *Real World Construction Scheduling* by Jim Coombes is available through Construction Planning and Management, P.O. Box 47355, Seattle, WA 98146, (206) 935-9393.

Levels

Assemble the following:

- 6″ torpedo level
- 24–32 inch hand level
- line level (with string that can be tied to chairs)
- water or hydro-level (add coloring to increase visibility)
- builder's level or transit level
- laser level
- theodolite (computerized)

Demonstrate the proper use of each leveling device. Be sure to highlight the jobs each one is best suited to perform. The small easy-to-carry torpedo level can fit into areas that another level cannot. The hand level increases the accuracy over a longer surface. A straightedge can be used to enhance the range of use. Show how to check a level for accuracy (a chalkboard works). The string line is quite inexpensive, but useful in some applications. Using two chairs, demonstrate the shortcomings of the string line. The possible sag in the line necessitates pulling the line very tight. Secure posts are needed. Longer distances can be covered than with hand levels, however.

The often underrated water level is worthy of an extended class of its own. Besides the basic garden hose with clear plastic tubes attached, there are also several models from various manufacturers that contain their own water reservoir, 50 feet of tubing, and are used by one person. The long tube allows a 100-foot diameter work area without moving the level. They can be used in places, such as around corners, that a laser could not. Water levels take little set-up time and no electricity. Care must be taken to avoid spilling and damage to the tubing. Cart wheels rolling over or people standing on the hose is a frustration when using them.

The technological advancements in laser levels has been great. These labor-saving devices are available at costs to fit almost any budget. They are direct electrical plug-in or rechargeable battery powered. Some are leveled manually, while others are self-leveling. The smallest have been developed for acoustical ceiling installation. Most emit a light visible to the eye and some now use sound waves. We usually think horizontal when using a level, but like the reliable hand level, many models can be used to plumb as well. The time saved plumbing elevator shafts, the height of a building, or walls their entire length pays for the investment. There are numerous extras that can be purchased for specialty applications: wall mount brackets, racheting clamps, reflective targets, and prisms that allow turning a 90-degree corner for layout of intersecting walls. The leveling functions lasers perform often exceed the building level (plus it only takes one person to do the task). More is found on this topic in Chapter 5.

Using the builder's level, demonstrate proper handling, storage, and set-up. If there is a transit available, compare the similar functions and the superior range of the transit. The transit will be covered in Chapter 5 and a complete transit course should be developed to suit your training facility and surrounding space. Basic functions and varied uses are the purpose of this demo. Angles, triangles, degrees, minutes, and seconds are also covered in another class — this is an excellent time to mention them. If there is any possibility of getting a computerized theodolite for a class, please do so. This instrument is most generally used by surveyors, but is being used increasingly more by carpenters. Perhaps a surveyor in your area could be a guest speaker to give the class a background on surveying and an overview of their job. The hubs established by the surveying crew become the cornerstone of the carpenter's layout.

Allow the students to try out the different levels. Have a group discussion using situations and choices.

The Engineer's Rod

The engineer's rod has readings in feet and decimal parts of a foot. The large numbers on the rod are feet (often in red or orange). Each top edge of the black line by the foot marker is the correct foot value. The numbers between the feet are tenths of a foot. Each pointed line by the number is a tenth of a foot. The top and bottom of each black line is a hundredth of a foot. When adding and subtracting elevations, using tenths and hundredths is a lot easier than inches and eighths. Practice reading the rod so that you do not need to count the hundredths. Focus on one tenths segment. All of the others are the same.

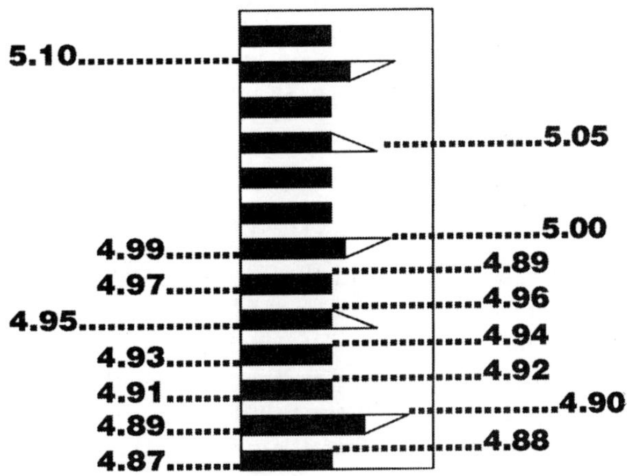

The longest points are tenths of a foot. The shorter points are five hundredths (.05). All points are counted by five hundredths. The rod is read from the bottom up. Any reading that falls midway on a black bar or white space would be a thousandth (.001) of a foot.

Converting Inches to Decimals

There are 8/8ths in an inch. Since there are 12 inches in a foot, there are 98/8″ in a foot. There are 100/100ths in a foot on the engineer's rod (1.00).

The following is true because each equals 1 foot: 96/8 = 100/100. When reading the engineer's rod remember that **0.01′ is about 1/8″.** This will help keep elevations in proportion. A difference of 0.09′ might not seem like much, but it is over 1 1/16″; 0.21′ is more than 2 1/2″.

The easiest way to remember the conversions without memorizing a table or computing the decimals every time is: **8–9–8.** Below are the decimal values for inches in a foot and their rounded-off values. Notice the pattern.

1″	.08333	.08	8
2″	.16666	.17	9
3″	.25	.25	8
4″	.33333	.33	8
5″	.41666	.42	9
6″	.50	.50	8
7″	.58333	.58	8
8″	.66666	.67	9
9″	.75	.75	8
10″	.83333	.83	8
11″	.91666	.92	9
12″	1.00	1.00	8

To convert inches into decimals, divide by 12. One inch is 1/12 of a foot. One divided by 12 equals .08333. This is rounded off to .08.

To change a decimal of a foot into inches, multiply by 12. Be sure to place the decimal point in the right place. An example is .42 × 12 = 5.04. There are 2 decimal places (tenths and hundredths) in the equation, so the answer will also have 2 decimal places.

Money Concept to Read Rod

Students can understand the principles of the foot divided into tenths and hundredths when an analogy with money is given. Referring to the full feet as dollars (the red numbers), tenths as dimes (the black numbers), and the bars for hundredths as pennies, helps the student relate to something that is both familiar and well-understood. The pointed portion of the hundredths is counting by fives and well represents a nickel.

This presentation also aids in the necessary math of adding and subtracting to find the height of instrument and elevations. I have found most people to be very receptive to this approach. In the beginning stages, the equivalent fractions can be limited to a hundredths, or about an eighth of an inch. As the class advances, more time can be devoted to converting decimals and inches.

Using the money concept, students can also see the decimal representative of the foot quite easily. When they read .25, they can understand that it is 1/4 of a foot. One quarter of a foot is 3 inches. Three quarters (.75) is readily changed to 9 inches.

Sketching Exercises

This classroom activity aids the students' understanding of blueprints and enhances their ability to conceptualize in three dimensions. Taking top, front, and right-side views of wooden blocks and drawing them as isometrics brings flat lines to life. After several blocks the process is reversed and the students change the isometric drawing to the three views. The drawing can be done without the blocks, but many students profit from touching and turning the block, to view it from all sides. Having the three dimensional block also affirms the need to show more than one view. An analogy of plans and elevations is used.

Pass out copies of views and grid paper. Have pencils and straightedges required or on hand. Show the students the block. Label the top, front, and right side. Rotate the block to show the three views. Mark the starting point and the grid on the block (3/4″ grid squares). Using either an overhead projector or board, begin drawing the figure. Pass the block around for students to examine. After you have completed the isometric, walk through the class and check on their progress. This activity might be very frustrating for students who cannot picture the results. Encourage and help them to understand that this is a job skill they will use. Carpenters need to see a footing and stem wall from a *flat* foundation plan.

Continue with all of the blocks. As the students gain confidence, they will proceed more quickly than your drawings. They will attempt to draw without seeing the block. This is what you want to happen.

Exercises

Sketch isometrics of problems 1–6:

1.

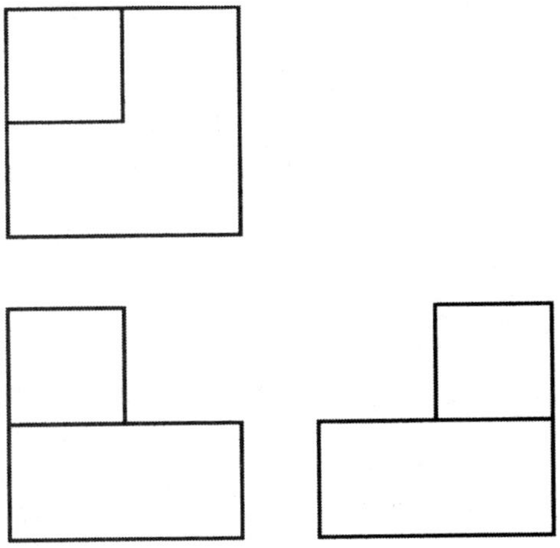

2.

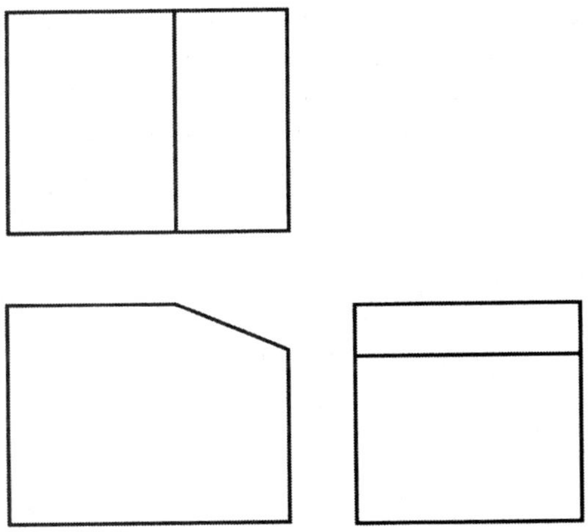

3.

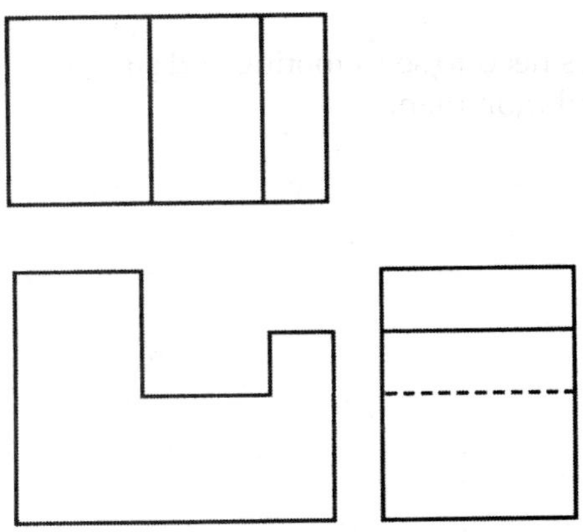

4.

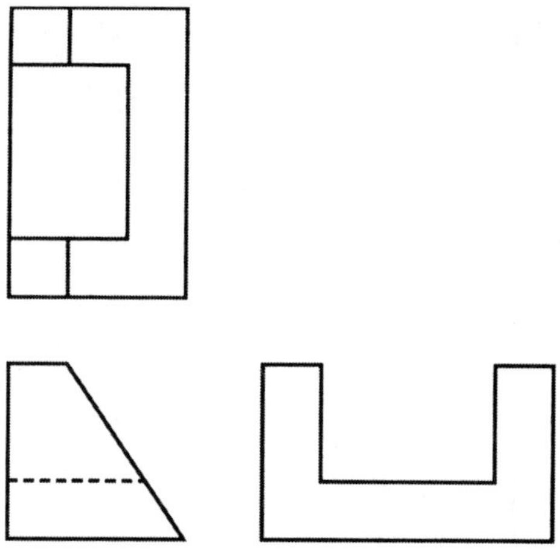

5.

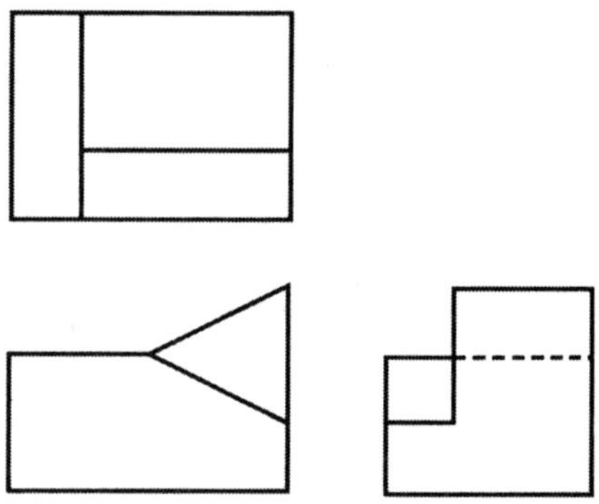

6.

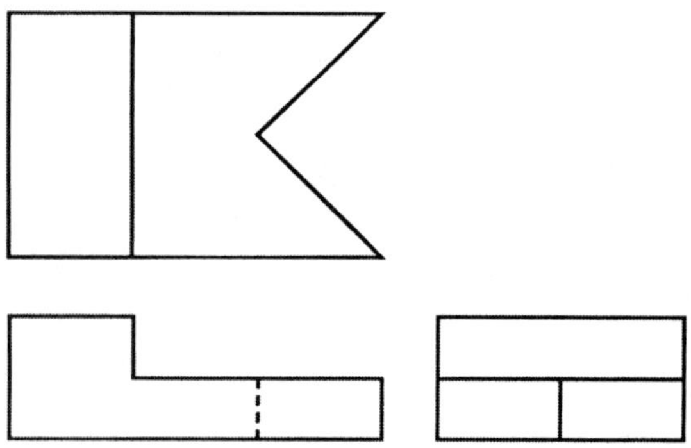

7. Sketch multiviews (top, front, and right side) of problems 7–9:

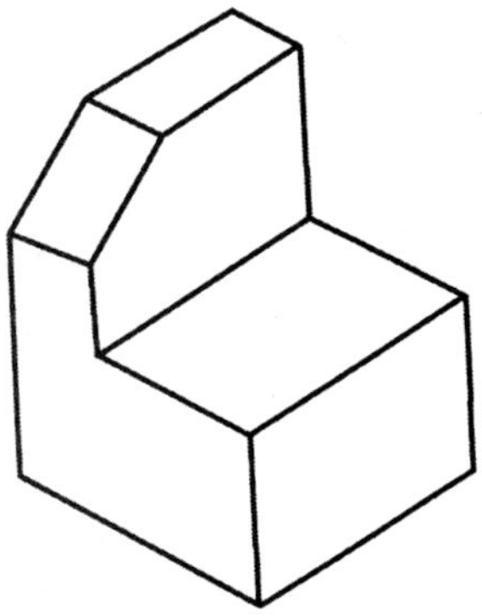

8.

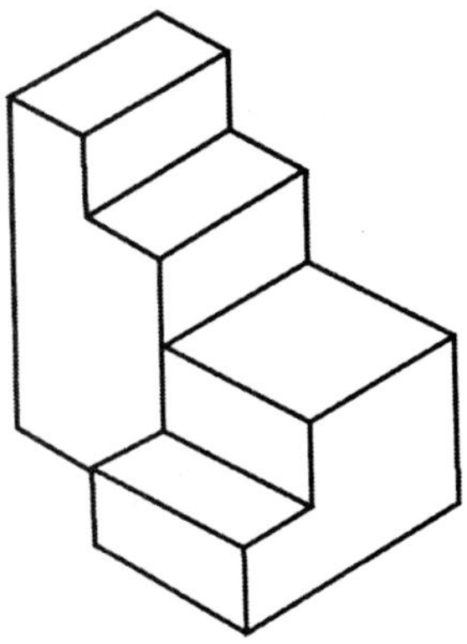

9.

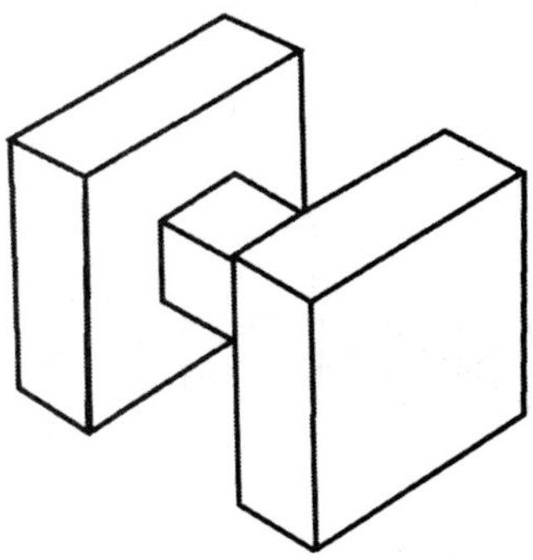

Sketching Answer Sheet

1.

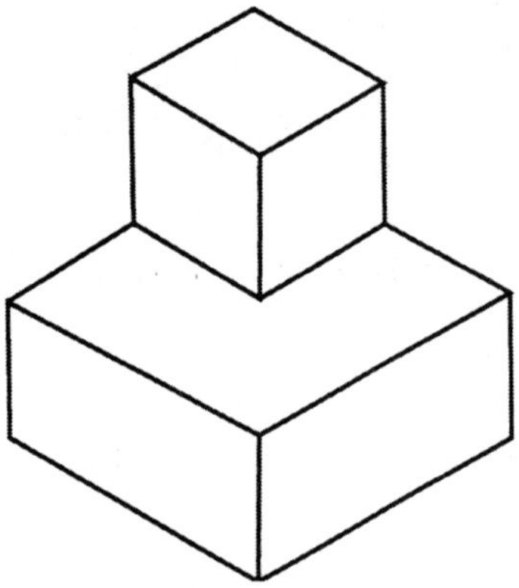

2.

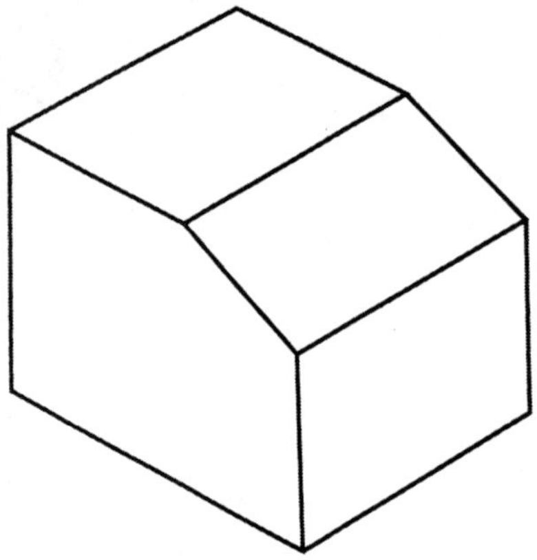

3.

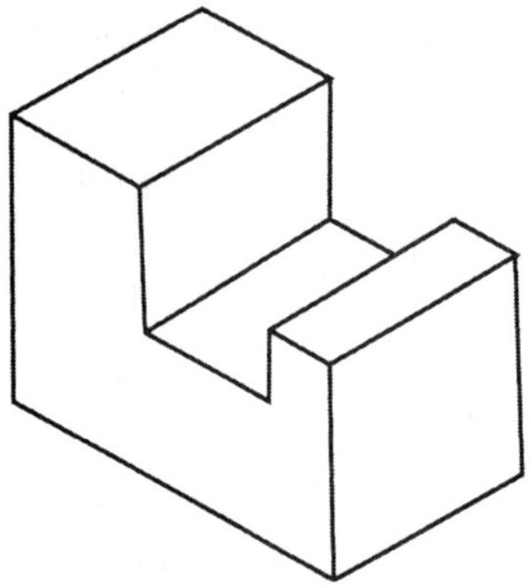

4.

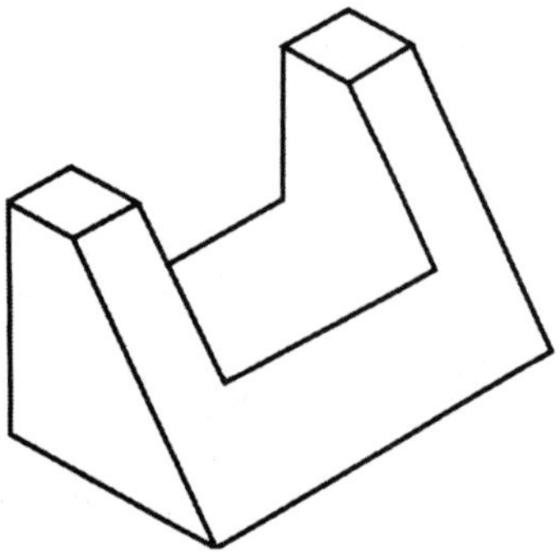

5.

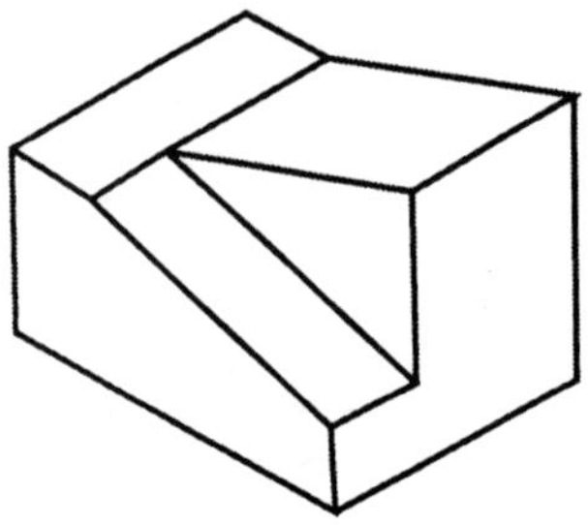

6.

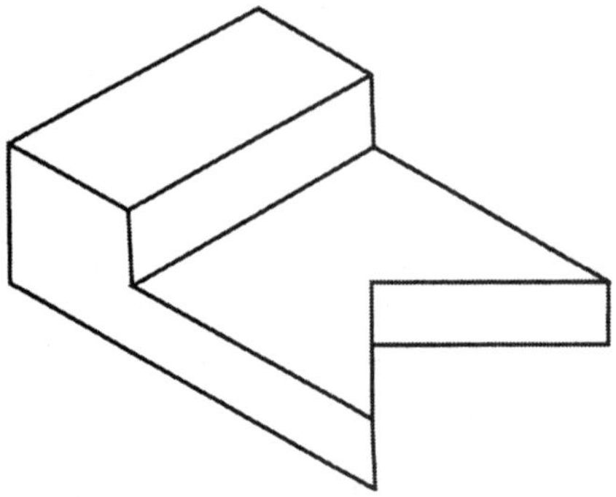

7.

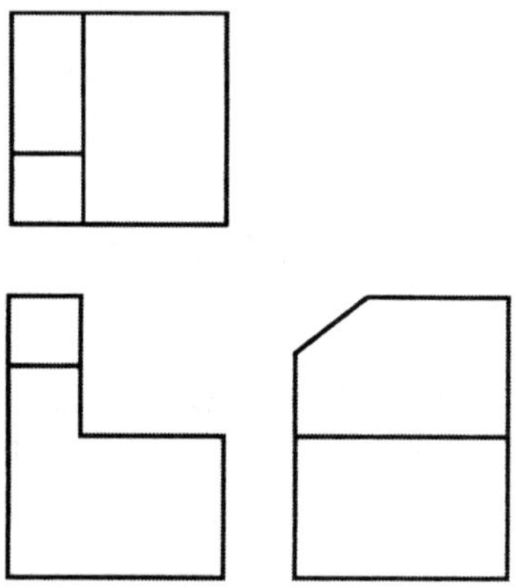

8.

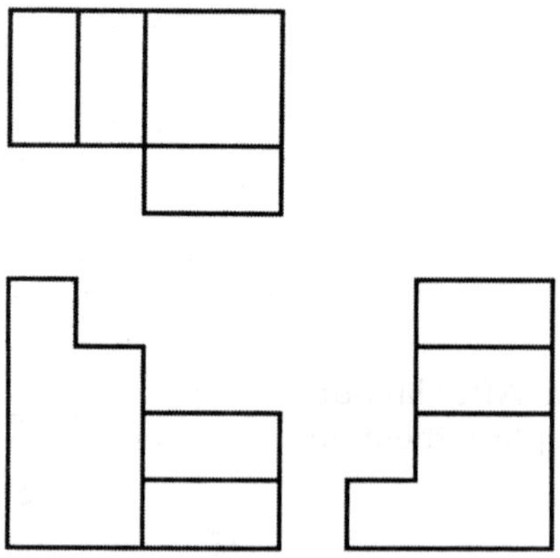

9.

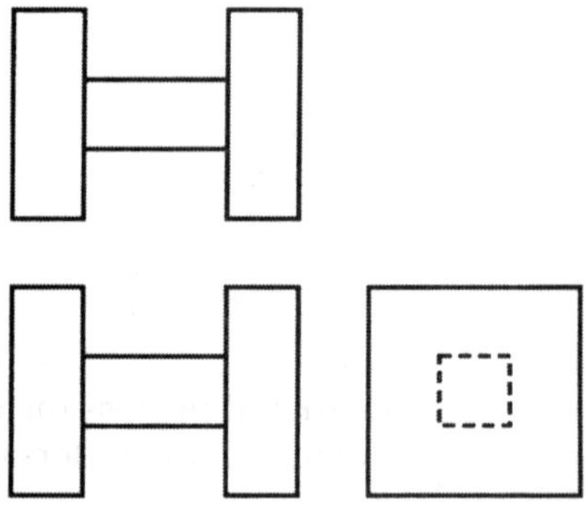

Roof Framing

Review the right triangle with the introduction of roof vocabulary. The sides become the rise and the run, while the hypotenuse is the rafter length. The accompanying sketches illustrate this. A combination of lecture, drawings (refer to the illustrations in the text, use handouts, the overhead, or on the board), demonstration, and hands-on practice make roof framing go from theory to reality. When teaching layout of common rafters, try to show four different ways of obtaining the length. This helps the student see the relationship of the terms of the roof and to select the method that is the easiest for him or her to apply.

The mock-up on page 180 can be used for both gable and hip roofs. It is a simple base with 2 × 4 plates, a 2 × 2 ridge and a plywood back with proper rafter slope drawn. This display takes minimal space (2′ × 4′ × 6′ tall) and allows the students to cut and check rafters using short lengths of lumber (2′ run).

Simplicity must be enforced. The theory and its application will stay with the student long after the class is over. Providing the students with the needed tools for problem-solving should be the goal of education. Some buildings that they may work on will have rafters and ridges made of 2× material, the next might be built with glulams or TJIs. The student/carpenter must make the transition by knowing the theoretical lengths and making necessary adjustments. The basics of plumb, level, and square will never become obsolete, although many of the materials used in construction will. Even the dimensions will change as we adopt the metric system. The ability to apply knowledge in new situations is a more valuable asset for the carpenter than being able to hammer 30 nails in a minute.

A class on common rafters can easily be divided into shorter segments to give the students a chance to digest and process the data. After layout, have the students cut the rafters and place them on the mock-up.

Building a Rafter Mock-Up Table

Material list:

3/4 plywood	1-4′ × 8′ sheet (4′ × 6′ back and 4′ × 2′ top)
1/2 plywood	1-4′ × 8′ sheet (front and sides of base)
2 × 2	1-3′ piece (ridge)
2 × 4	1-8′ piece (top place for rafter layout replaceable)
2 × 4 or other	14-4′ pieces (for base framing)
nails or screws	(as needed for assembling)
thin-line permanent marker	(lay out lines and labeling)

Build the base with any sound framing materials in the shop. These do not have to be 2 × 4′; 2 × 2 or 2 × 3 can be used for economy. The framed size of the base is 3′-8″ wide, 1′-10 3/4″ deep, and 2′-11 1/4″ tall.

After framing is complete, wrap the sides and front with 1/2″ plywood. Apply the 6′ × 4′ back of 3/4″ plywood. Use the remaining 2′ × 4′ piece of 3/4″ plywood for the top (there should be a 1 1/2″ overhang in the front and on both sides from the framing with the back and top flush).

Screw the top place on for easy replacement. Use permanent marker to lay out rafters on the plate. Chalk the vertical centerline on the back plywood. Measure up minimum thickness of wood above the seat cut (plans use 2 1/2″). Snap chalk lines to centerline for unit rise from 2″-12″. (These lines may be sprayed with clear lacquer or drawn over with a marker.)

Nail ridge board in place. Label unit rise and continue rafter lines to center.

Plans to Build

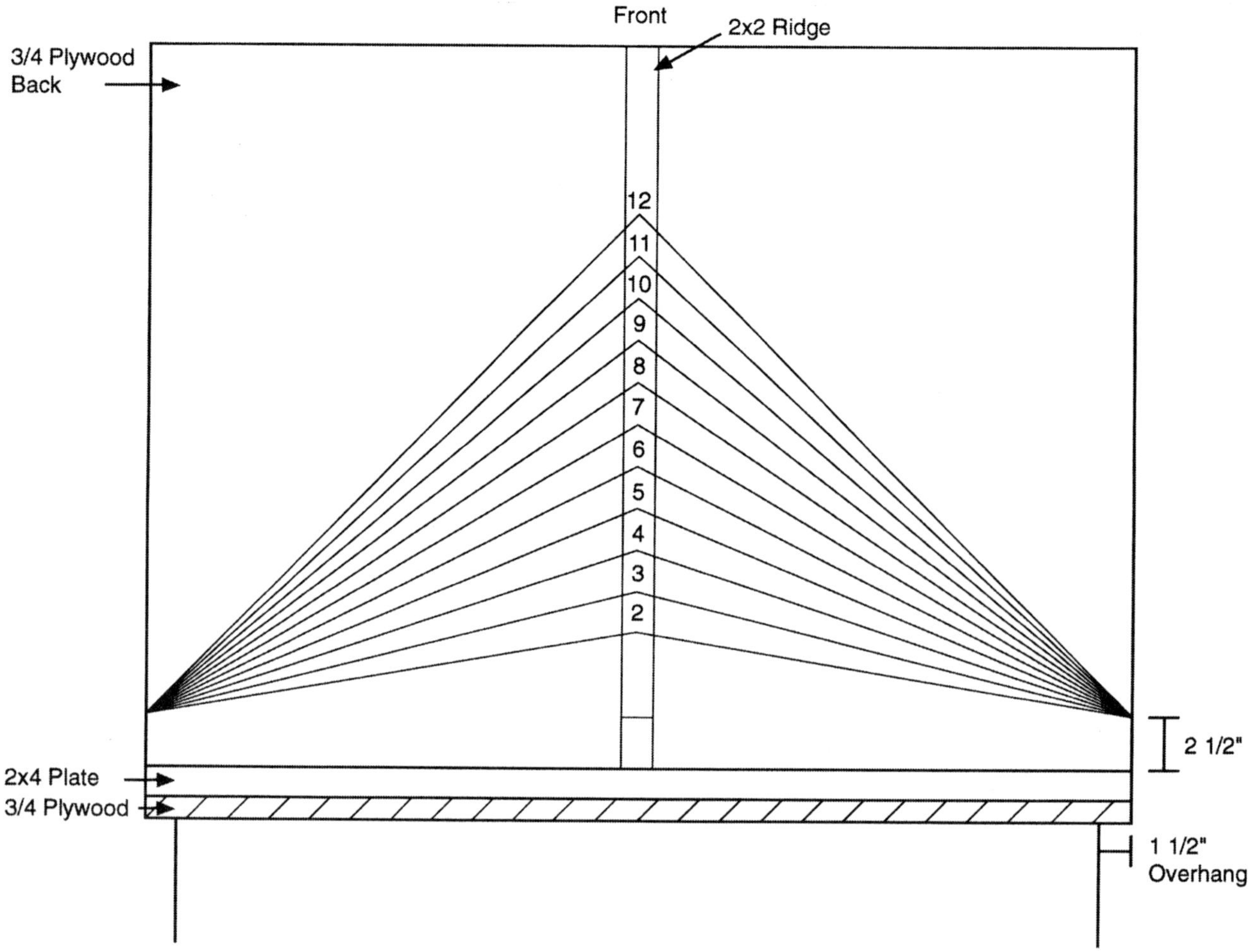

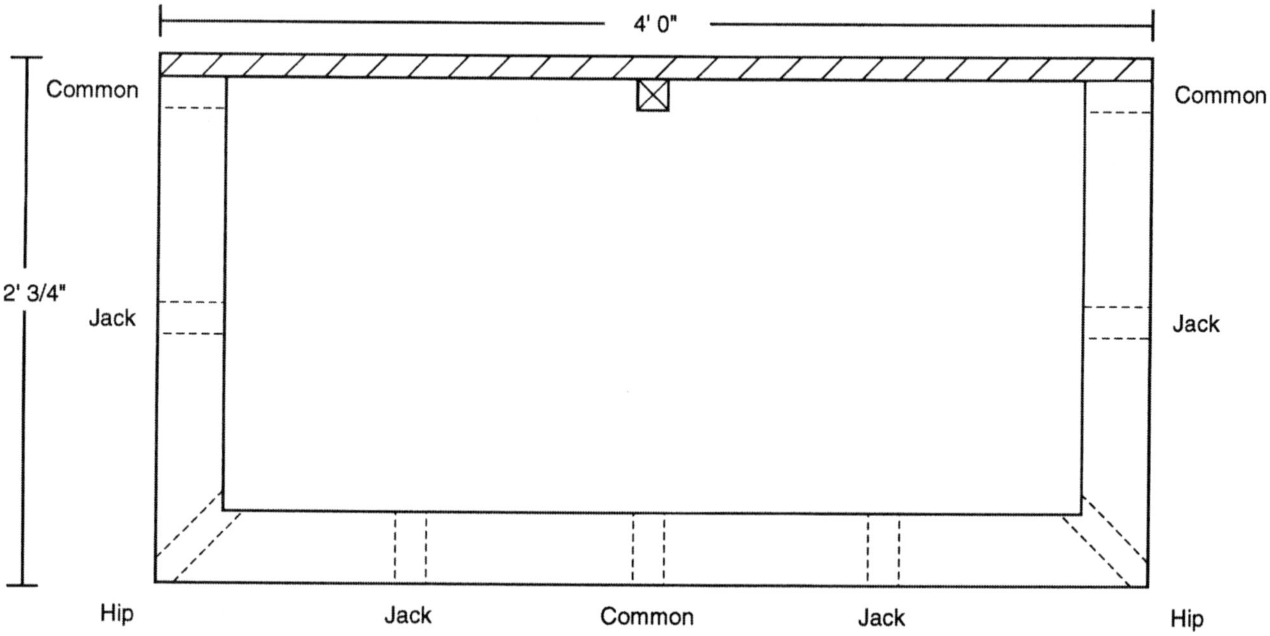

Top

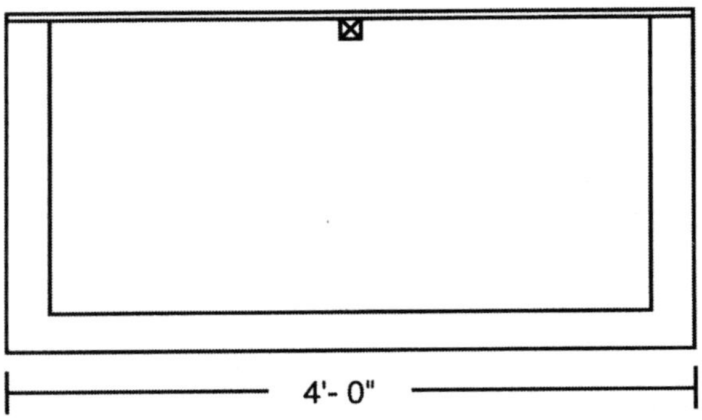
4'- 0"

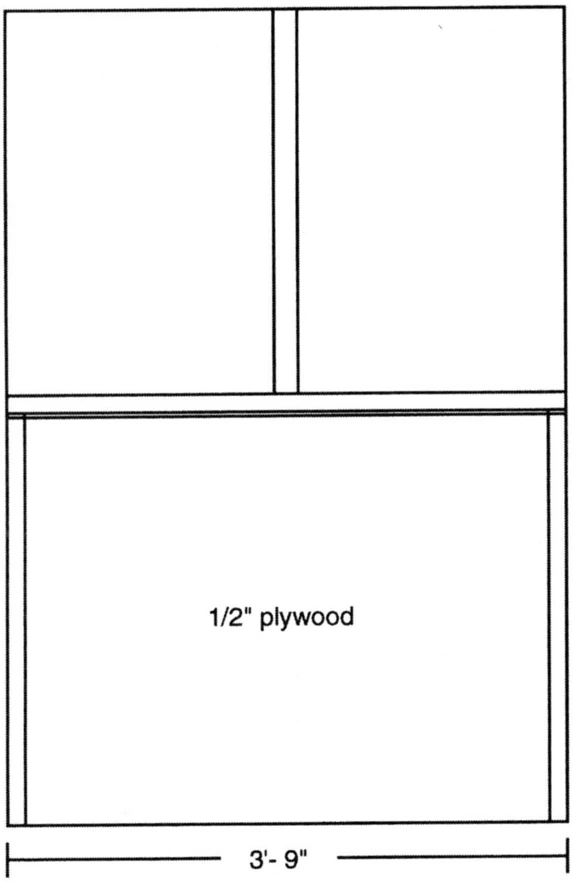
1/2" plywood
3'- 9"

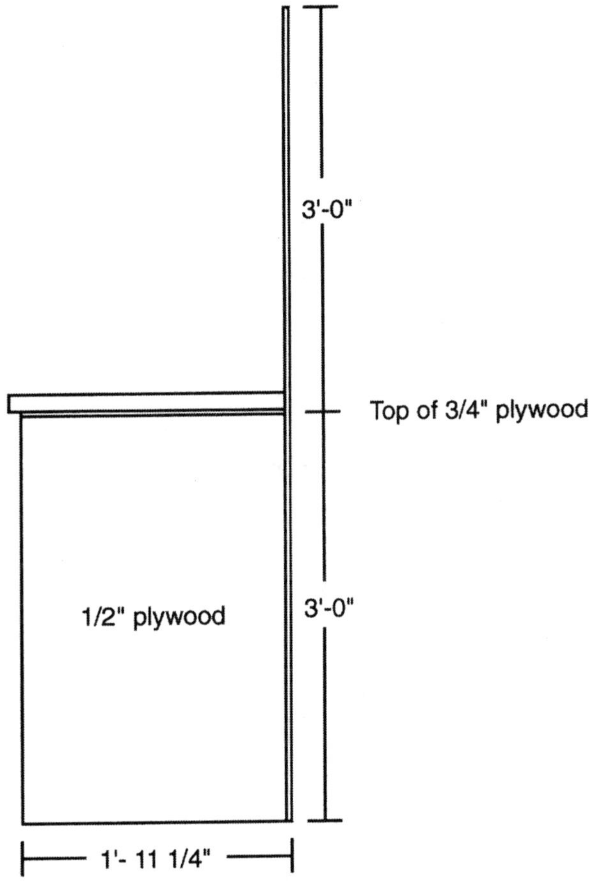
3'-0"
Top of 3/4" plywood
1/2" plywood
3'-0"
1'- 11 1/4"

Common Rafter Practice Exercises

Questions 1–5 refer to this building.

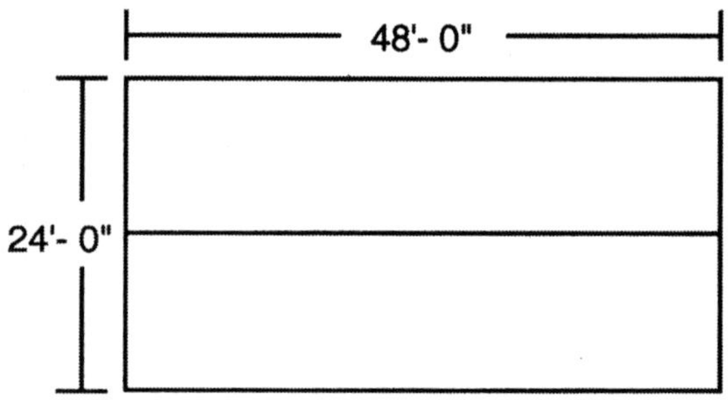

1. What is the span of the building? ______________
2. What is the run? ______________
3. If the unit rise is 5″, what is the total rise? ______________
4. What is the length of common rafter? ______________
5. How many rafters are needed if they are framed 2′ OC? ______________

Questions 6–10 refer to this building.

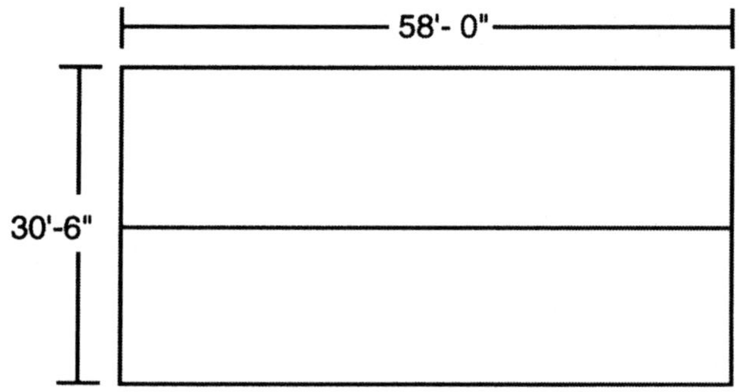

6. What is the run? ______________
7. What is the length of common rafter with a 7-in-12 slope? ______________
8. How many rafters are needed at 2′ OC? ______________
9. With a 12″ overhang, how long must the lumber be to the nearest foot? ______________
10. How many board feet are needed for the roof if the rafters are cut from 2 × 12 20′ lumber? ______________

Hip Roof Practice Exercises

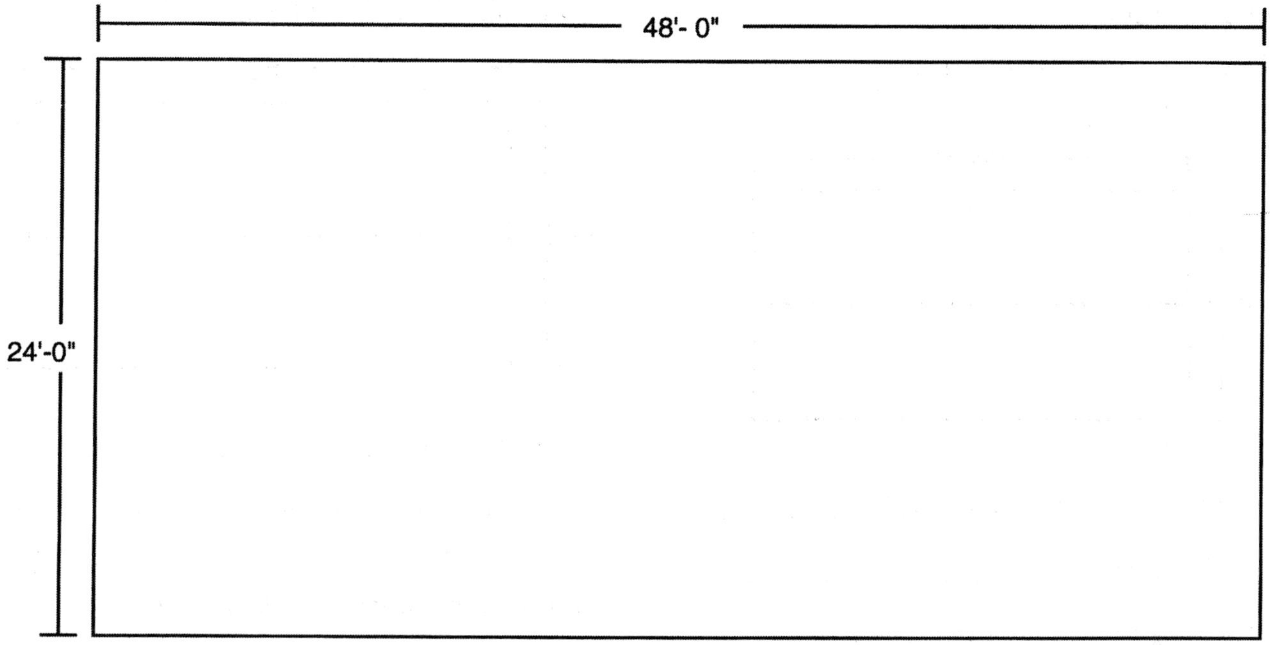

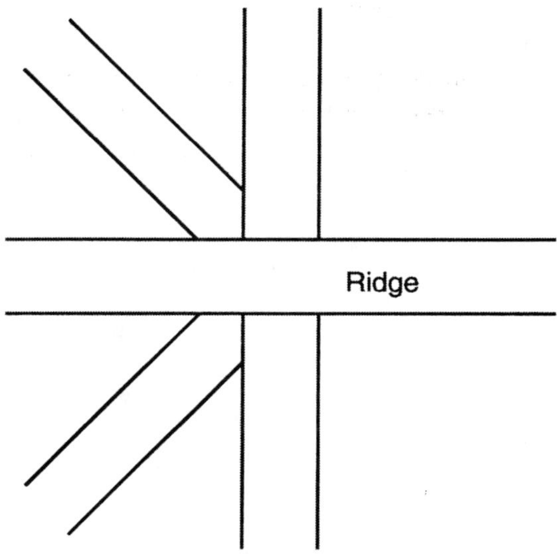

2 x 6 Rafters 2' O.C.
2 x 8 Ridge
1/3" Pitch

1. Lay out all rafters.
2. What is the theoretical length of the ridge? ________________
3. What is the actual length of the ridge? ________________
4. What is the theoretical length of the common rafters? ________________
5. What is the theoretical length of the hip rafters? ________________
6. What is the difference in length of the jack rafters? ________________

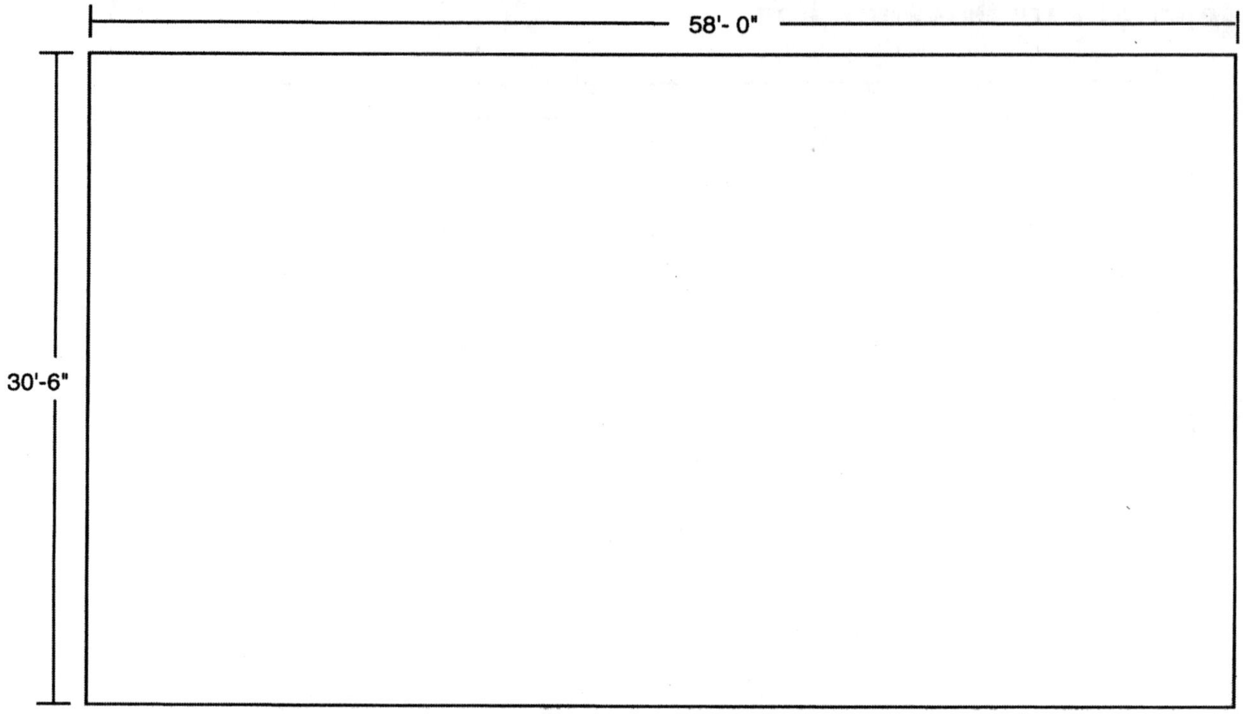

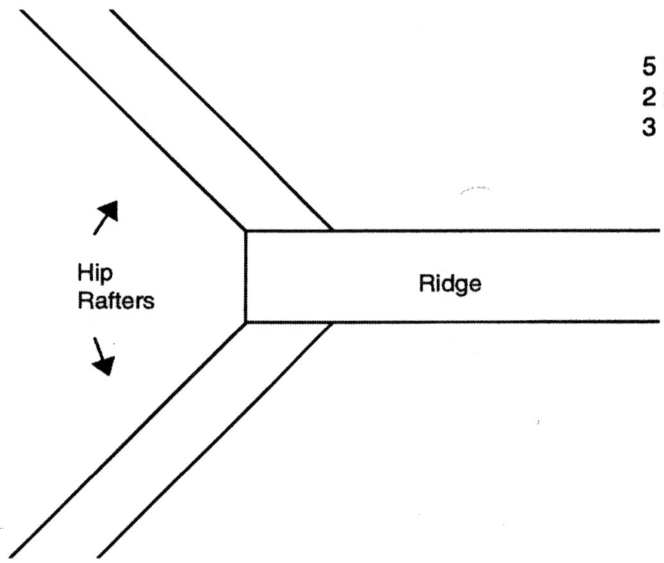

5 in 12 Slope
2 x 8 Rafters 2' o.c.
3 1/2" Glulam Ridge

1. Lay out all rafters.
2. What is the theoretical length of the ridge? __________
3. What is the actual length of the ridge? __________
4. What is the theoretical length of the common rafters? __________
5. How many common rafters are there? __________
6. What is the actual length of the hip rafters? __________

Picnic Table Plan

This picnic table is made from 2 × 4 and 2 × 6 material. When made with pressure treated wood, it will not rot or decay due to weather exposure or insect infestation. This will also add to the table weight.

Material list:

9-2 × 6-8′	Table top and seat
4-2 × 6-3′	Legs
2-2 × 6-52″	Seat support
2-2 × 4-26″	Table top support
2-2 × 4-39″	Brace
1-2 × 4-4′	Seat and table top cleats
12-3$^{1}/_{2}$″	Carriage bolts with nut and washer
70-3″	Deck screws

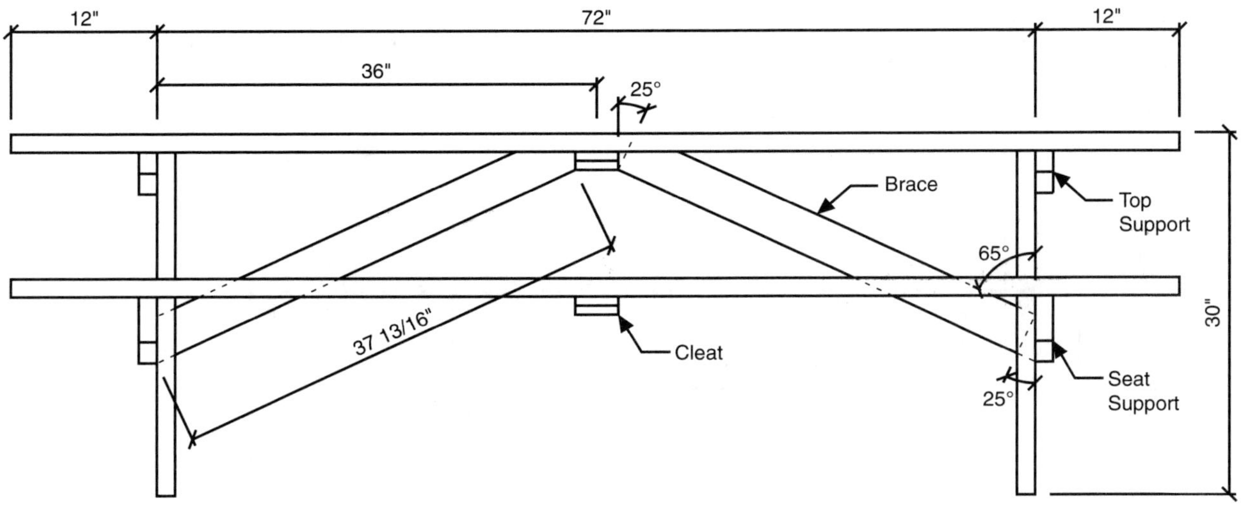

Side View

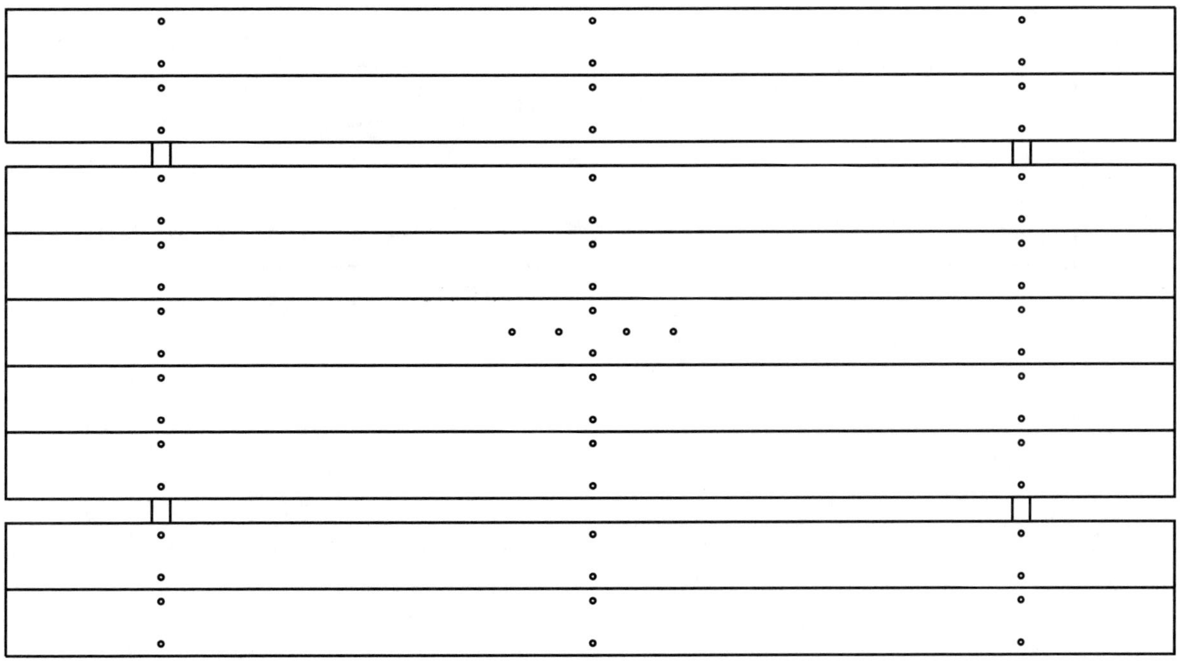
Top View

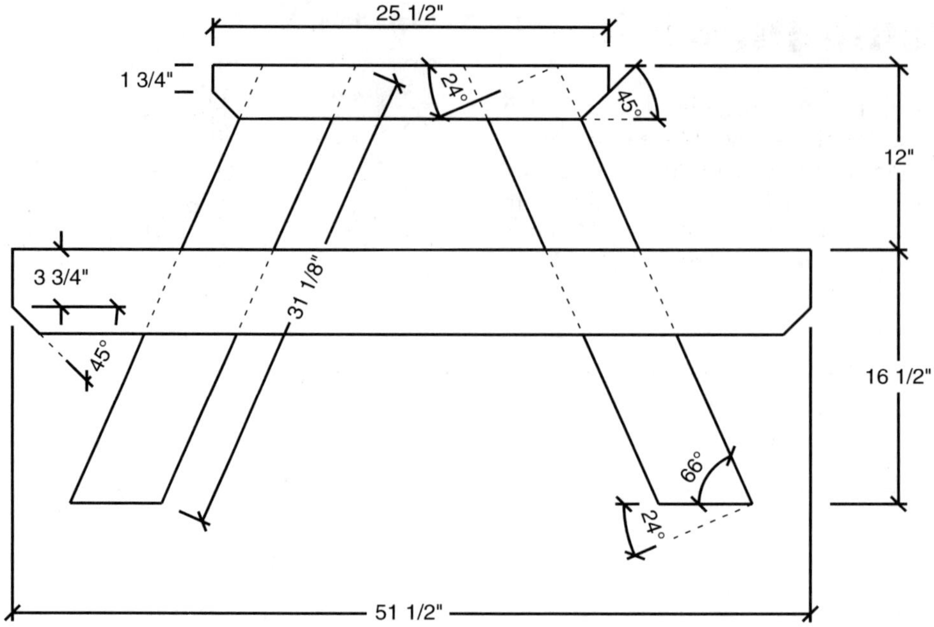

Framing Detail

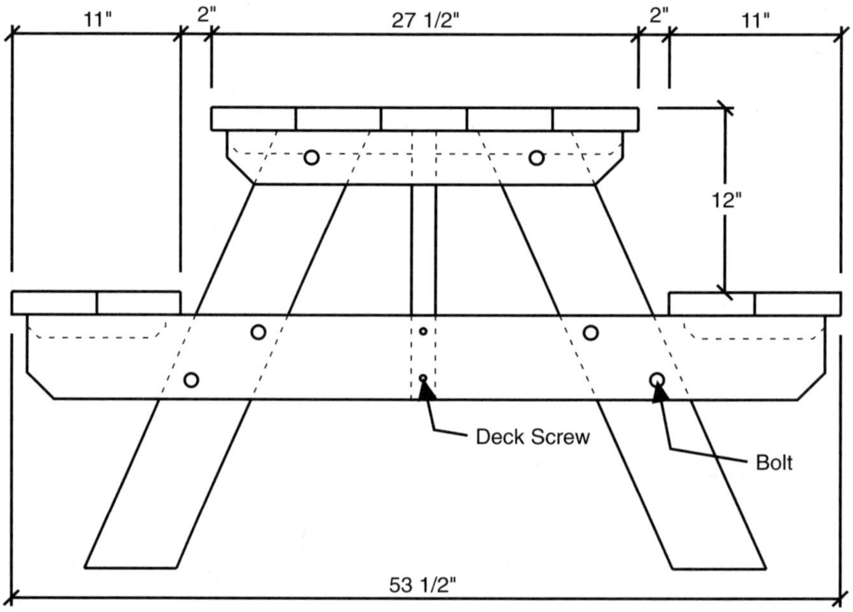

End View

Small Tool Bench Plans

This is an excellent small project that can be completed by inexperienced students. The entire bench is made out of 3/4-inch plywood. The layout will use square lines, tapered lines, radius corners, and a 2-in-12 slope. For the total beginner, some parts may be precut to save material. The small size makes storage and assembly areas minimal.

The variety of power tools used is only limited to the training facility. The table saw, skill saw, drill motor, jig saw, band saw, router, and sander could all be used during production. All joints are glued and nailed. The finished bench can also be painted, stained, or laminated, if desired.

The benches can be used at home (given for holiday gifts), sold at craft or trade fairs, donated, or used in class.

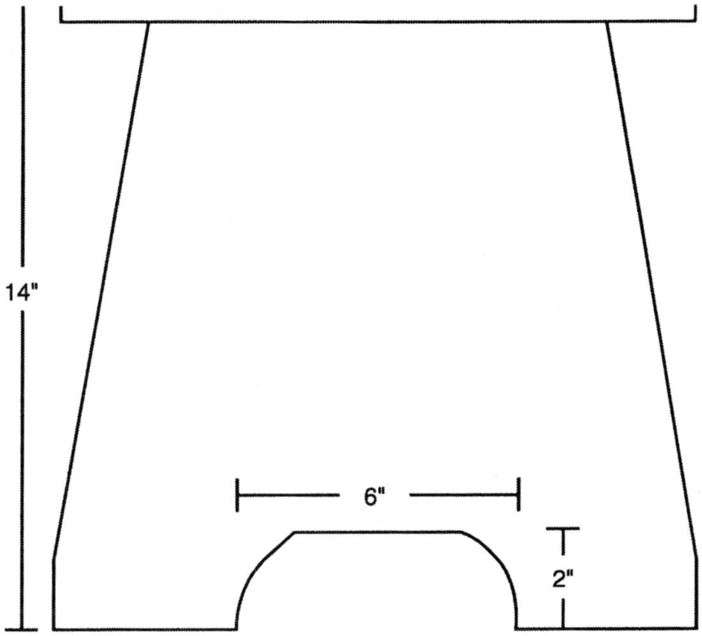

End View

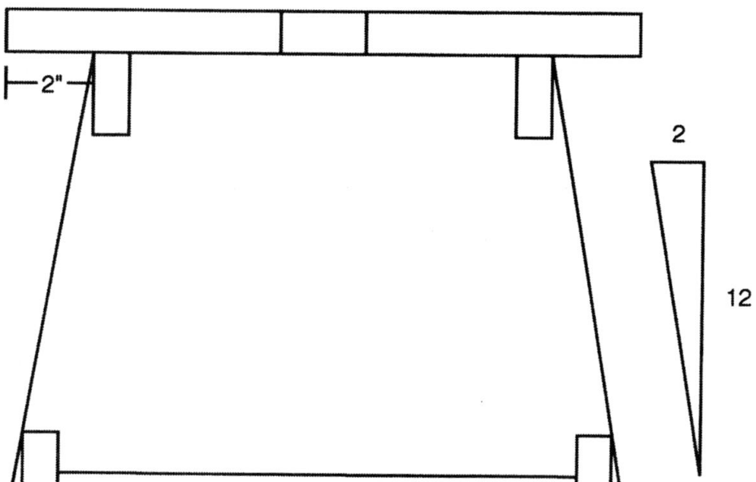

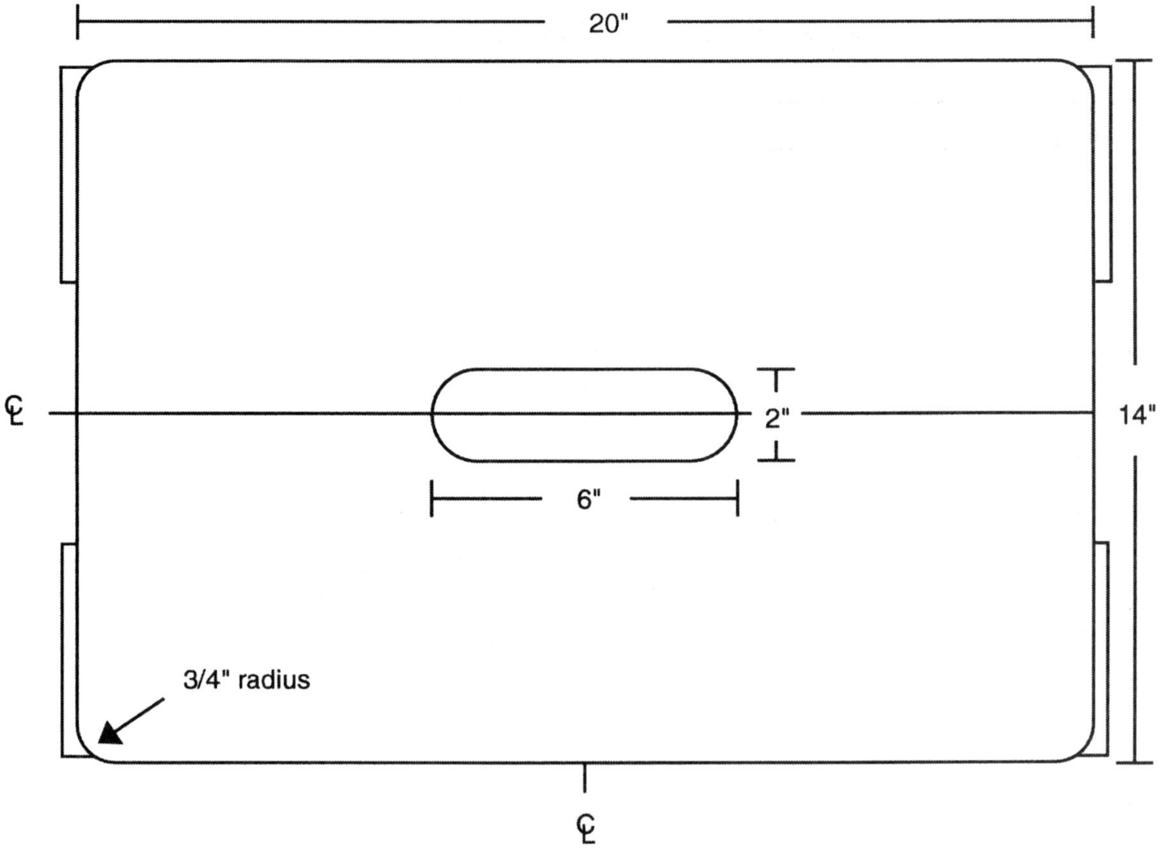

Top View
Bench

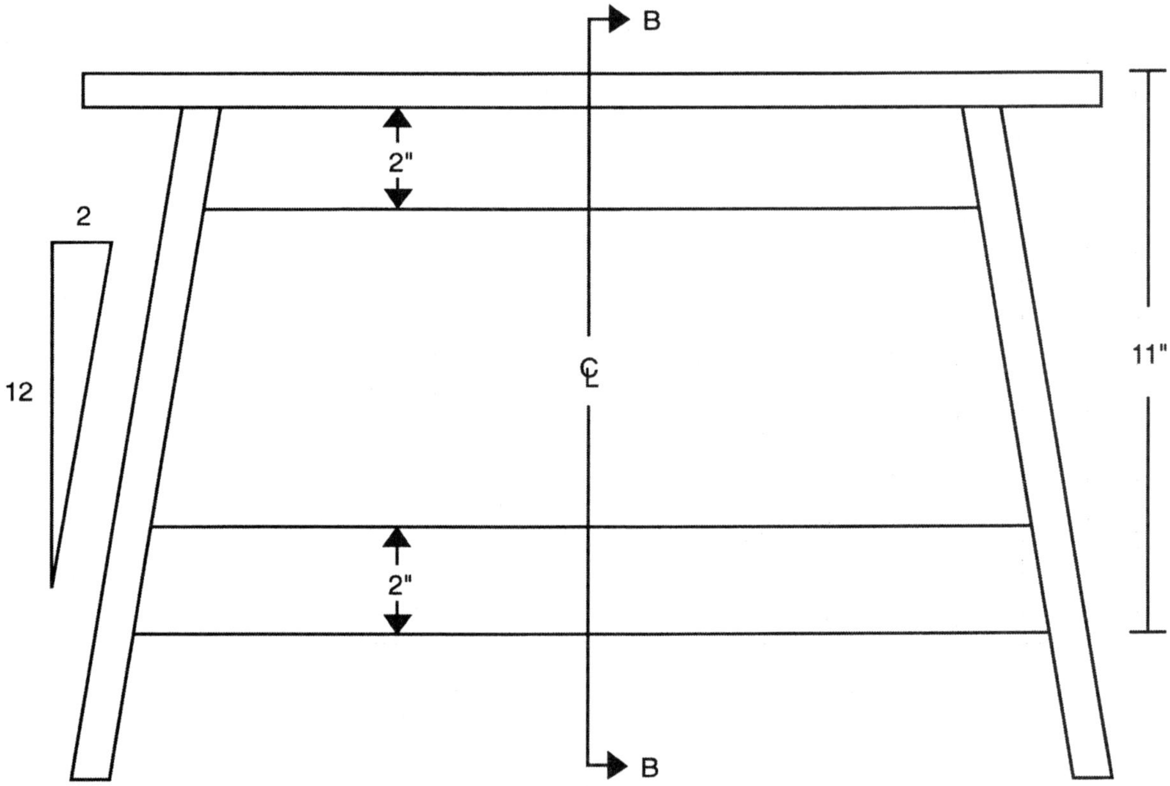

Front View

Measuring Station Plans

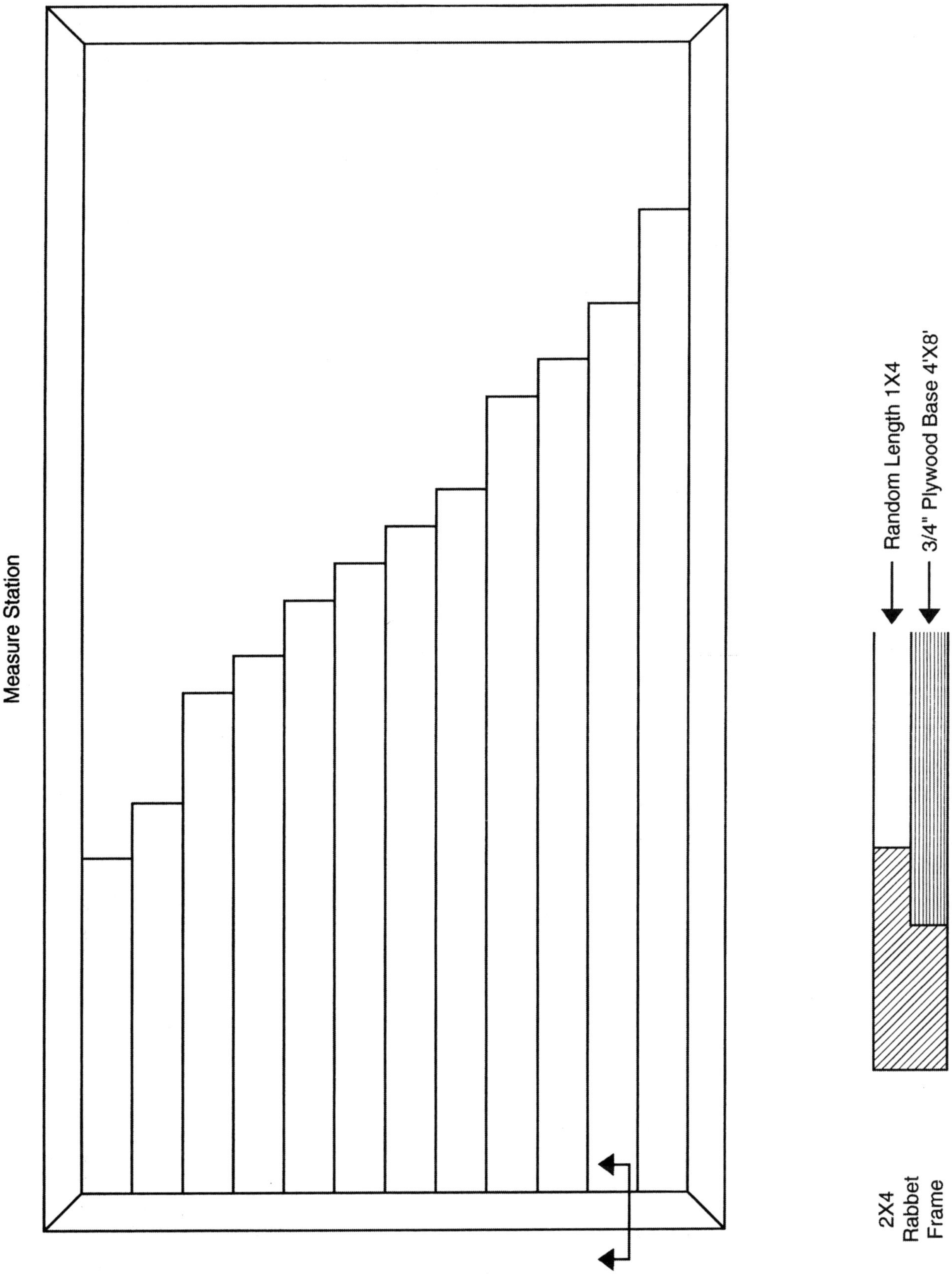

Nailing Station Plans

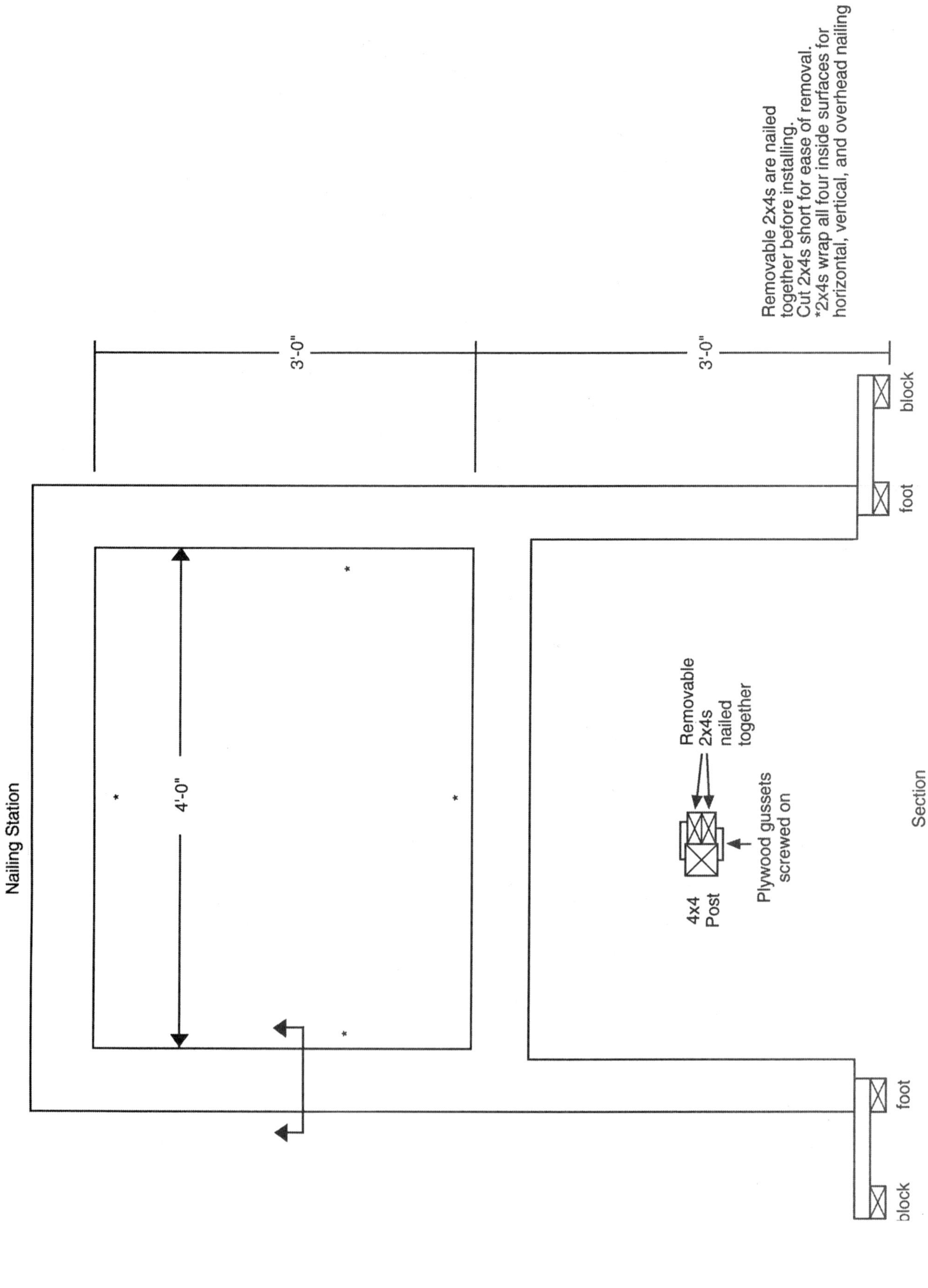

Portable Module Plans

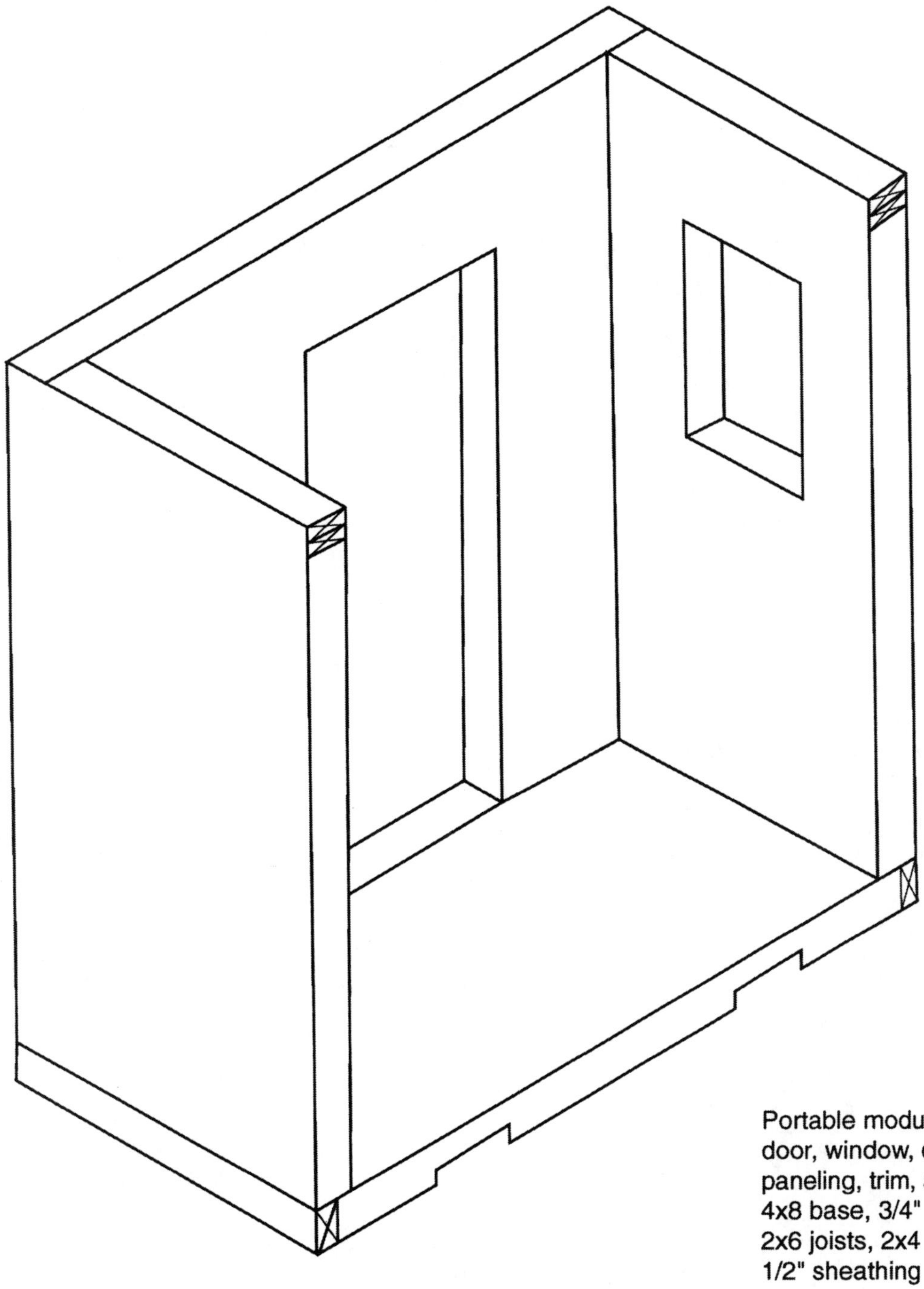

Portable module for door, window, drywall, paneling, trim, and flooring. 4x8 base, 3/4" ply over 2x6 joists, 2x4 walls, 1/2" sheathing

Roof Rafter Illustration

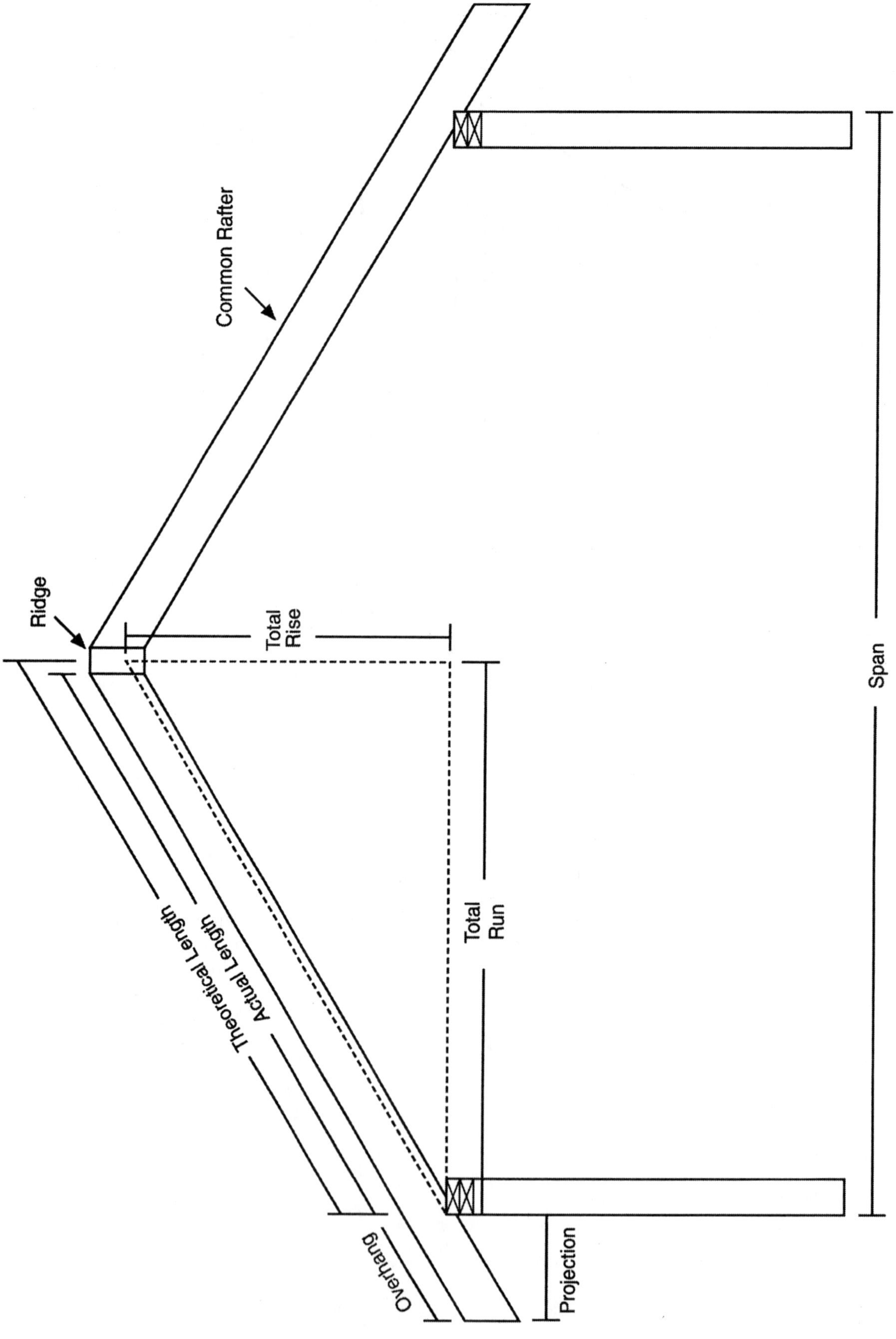

Trim Practice Board

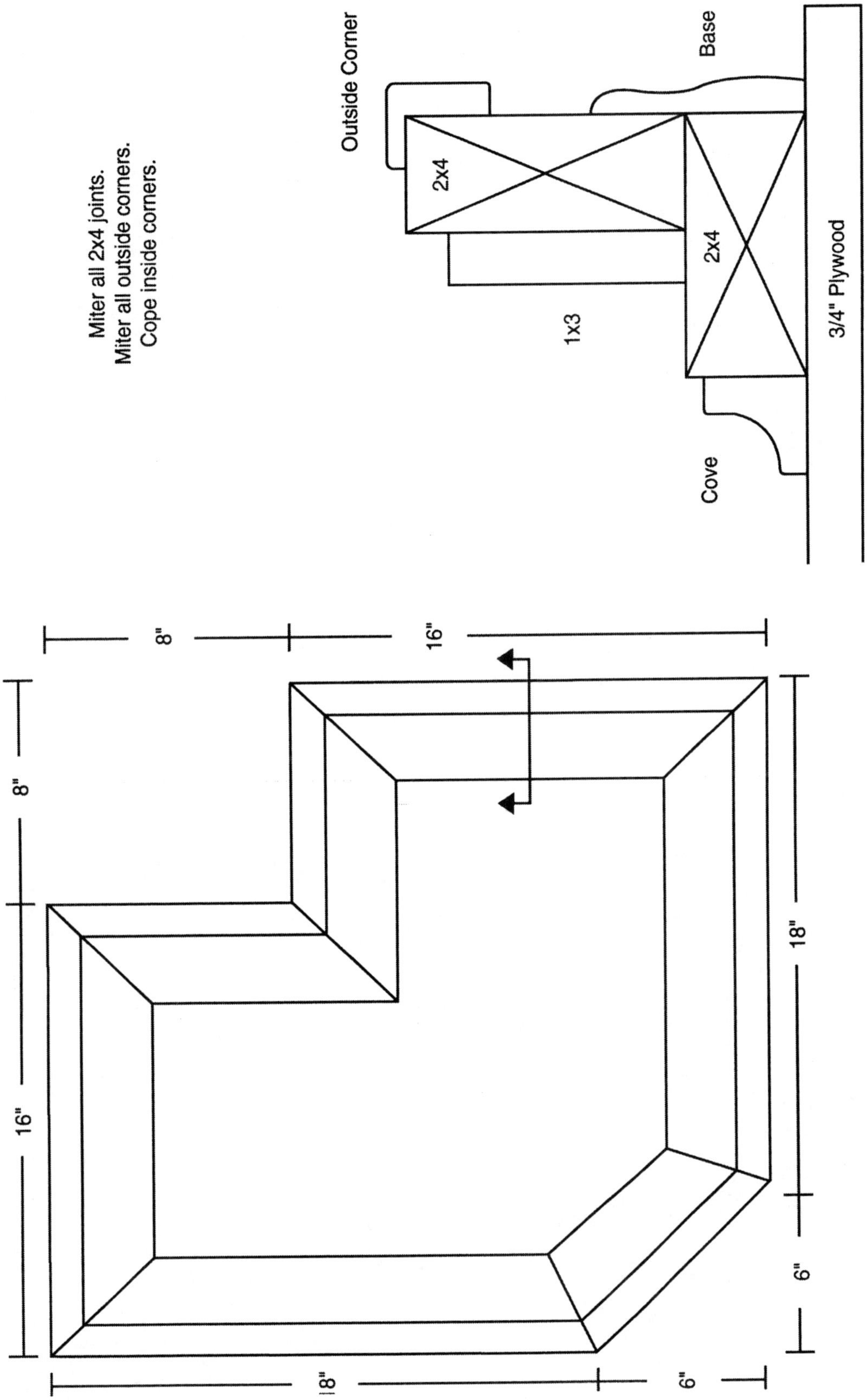

Trim Practice Board

Answers to Math Review Questions

Working with Feet and Inches

1. 12′
2. 63′
3. 10′
4. 8′
5. 7′
6. 144′
7. 144″
8. 100″
9. 204″
10. 187″
11. 144″
12. 53″
13. 8-10″
14. 11′-1″
15. 187′-2″
16. 113′-0″
17. 16′-2″
18. 8′-2″
19. 3′-10″
20. 16′-0″

Linear Measurements

1. 6222 ft.
2. 84 ft.
3. 215 ft., $220.32
4. 112 ft.
5. 576 ft.
6. 26.67 (27 rolls)
7. 16 pieces

Fractions

1. 2/3
2. 5/8
3. 3/4
4. 7/8
5. 1 5/8
6. 12 1/12
7. 4 1/2
8. 12
9. 4/16
10. 6/16
11. 8/16
12. 14/16
13. 3/4
14. 15/16
15. 1 1/2
16. 15/16
17. 1 19/32
18. 35/48
19. 66 1/4
20. 89 11/16
21. 5 49/64

Subtracting Fractions

1. 1/4
2. 3/8
3. 1/2
4. 1/8
5. 7/16
6. 3/16
7. 5/8
8. 11/32
9. 1 13/16
10. 3 3/8
11. 6 3/4
12. 2 5/16
13. 3 1/8
14. 12 1/4
15. 1 1/4
16. 3 9/16
17. 13 1/8
18. 8 3/4
19. 10 15/16
20. 7 31/32
21. 9/16

Multiplying and Dividing Fractions

1. 3/8
2. 1/16
3. 15/128
4. 12
5. 3/128
6. 3/8
7. 1 55/64
8. 3/224
9. 37 37/64
10. 168 3/4
11. 3/4
12. 1/2
13. 13/20
14. 9 11/13
15. 2 5/6
16. 8

Word Problems with Fractions

1. $24.09
2. 22 offices
3. 21 yards
4. 10.33 hrs
5. 4 1/2″
6. 7 1/8″
7. 119
8. $177,888
9. 15515.5
10. 21440
11. 1 3/4″
12. 1869.2′
13. 4316
14. 32 sheets
15. 237
16. 237
17. 95′
18. 6 1/4″

Board Feet Exercises

1. 80
2. 630
3. 896
4. 828
5. 170
6. 396
7. 196
8. 120
9. .61 (11/18)
10. 2.48 (2 69/144)
11. 75
12. 20
13. 260
14. 93.3 (93 1/3)
15. 17

Area of Plane Figures

1. 144 sq. ft.
2. 225 sq. ft.
3. 784 sq. in.
4. 12.25 sq. ft.
5. 3.06 sq. ft.
6. 19.5 sq. ft.
7. 600.25 sq. ft.
8. 1314.0625 sq. ft.
9. 30.25 sq. ft.
10. 14.0625 sq. ft.
11. 35 sq. ft.
12. 88 sq. ft.
13. 136 sq. in.
14. 6.20 sq. ft.
15. 14.13 sq. ft.
16. 29.81 sq. ft.
17. 1031.25 sq. ft.
18. 302.8125 sq. ft.
19. 246.75 sq. ft.
20. 277.5 sq. ft.
21. 113.10 sq. ft.
22. 314.16 sq. in
23. 254.47 sq. in.
24. 615.75 sq. ft.
25. 38.48 sq. ft.
26. 7.069 sq. ft.
27. 50.27 sq. ft.
28. 380.13 sq. in.
29. 254.47 sq. in.
30. 60 sq. ft.
31. 45.5 sq. in.
32. 187 sq. in.
33. 4.93 sq. ft.
34. 10.67 sq. ft.
35. 80.5 sq. ft.
36. 200 sq. in.
37. 9.92 sq. ft.
38. 145.17 sq. ft.
39. 286 sq. in.
40. 18.06 sq. ft.
41. 24.97 sq. ft.
42. 21 sq. in.
43. 40 sq. ft.
44. 20.72 sq. ft.
45. 17.99 sq. ft.
46. 156.25 sq. ft.
47. 722 sq. in.
48. 41.57 sq. in.
49. 93.53 sq. ft.
50. 28.87 sq. ft.
51. 162.83 sq. ft.
52. 1743.05 sq. in.
53. 62.00 sq. ft.
54. 3785.47 sq. ft.
55. 474.83 sq. ft.

Volume Word Problems

1. 572.4 cu. yd.
2. 25 1/3 cu. yd.
3. 45.04 ⇒ 45 cu. yd.
4. 23.93 ⇒ 34 cu. yd
5. $7075.00
6. 63.6 ⇒ 64 loads
7. 119.38 ft
8. 3 1/3 cu. yd
9. a. 760 sq. ft.
   b. 17.4 ⇒ 18 pieces
   c. 26.125 ⇒ 27 pieces
10. 17.87 cu. ft
11. 21.31 cu. ft.
12. $227.82
13. 13.26 ton

Angles

1. 120
2. 60
3. 30
4. 90
5. 60
6. 43
7. 25
8. 40
9. 10
10. 12
11. 30
12. 20
13. 48
14. 21
15. 40
16. 20
17. 204°-0′-20″
18. 119°-27′-12″
19. 29°-51′-5″
20. 91°-24′-53″
21. 5
22. 4′-30″
23. 3°-5′

A. 90
B. 120
C. 60
D. 72
E. 30
F. 140°-31′
G. 73°-20′
H. 286°-40′

A. 139
B. 41
C. 139
D. 124
E. 56
F. 124
G. 56
H. 56
I. 124
J. 56
K. 119
L. 61
M. 119
N. 119
O. 61
P. 61
Q. 119
R. 55
S. 40
T. 45

Common Rafter Exercises

1. 24′
2. 12′
3. 60″
4. 156″
5. 50
6. 15′-3″
7. 211.82
8. 60
9. 19′
10. 2400 bd ft.

Hip Roof Exercises

2. 24′
3. 24′-1 1/2″
4. 173.04″
5. 225.12″
6. 28 7/8″

Hip Roof Exercises

2. 27′-6″
3. 27′-8 13/16″
4. 198 1/4″
5. 30
6. 269 13/16″

Appendix A PowerPoint Presentations

Appendix A lists corresponding numbered PowerPoint Presentation titles.

1. **Measurement**

   English vs. Metric measure overview • Typical ruler • Colored markings and increments • Fractional divisions and names of each

2. **Square vs. Cubic Measure**

   Linear measure • Square measure • Cubic measure • Sample calculations and examples

3. **Framing Square Tables and Scales**

   Eighths, tenths, twelfths and sixteenths scales • Rafter tables • Octagon scale • Brace scale

4. **Achieving Level and Plumb with Simple Tools**

   Spirit level and the bubbles • How to adjust if bubble is touching a line • Exaggerated error from a 2-foot level • How to check a level for accuracy • Using a plumb bob

5. **Right Angle Overview**

   Need for right angles in construction • 3—4—5 Method • Pythagorean Theorem

6. **Lumber Sizes and Board Foot Measure**

   Overview of changes in lumber sizes • Board and dimension lumber sizes • Board foot measure and sample calculations

7. **Fastener Sizes**

   Nail size chart showing changes in increment of sizes • Screw sizes and gauges • Water level

8. **Locating a Building on a Lot**

   Plot plan information • Setback lines • Locating front building line • Locating building corners

9. **Adjusting a Rectangular Layout to Square**

   Overview of rectangle; equal sides and interior angles, importance of it • Determining the incorrect measurement • Variations in the solutions to improve squareness

10. **Floor System Components**

    Names and parts • Typical order of construction

11. **Installing a Sill**

    Spacing and bolt locations • Snapping a line on top of foundation • Aligning and marking sill • Measuring the hole locations from the edge • Drilling and installing snug against shims

12. **Sill Layout**

    OC concepts and necessity • Variations of OC; 12,16,19.2 & 24″ • First OC location and set back • Remaining OC locations • Layout of overlapped joists at the girder

13. **Joist Layout and Assembly**

    Snapping a line on the sill • Joists installed around an opening • Remaining joists and reinforcement • Rim joists, bridging and blocking • Sheathing installation

14. **Bridging Layout**

    Types of bridging • Wood cross bridging placement • Wood cross bridging layout and cutting

15. **Wall Component Identification**

    Wall component names and locations • Variations in corner and partition details • Variations in header details • Rough opening concepts

16. Wall Type Identification

Load-bearing walls • Load-bearing partitions • Non-load-bearing walls • Non-load-bearing partitions • Summation of differences and similarities

17. Wall Plate Layout Example

Layout of partition • Layout of opening centerlines • Layout of OC studs and cripple studs

18. Wall Assembly and Erection Example

Arrange and tack plates to deck • Layout without many details • Remove tack nails and separate plates • Assemble the components around an opening • Fill in remaining joists • Add doubled top plates • Raise and temporarily brace

19. Rafter and Roof Component Parts

Types of roof structures and styles • Types of rafters; common, hip, valley, and jacks • Names of rafter parts: ridge cut, bird's mouth, tail cuts • Projection vs. overhang

20. Unit Triangle and Rafter Theory

Rise, run, and length for a building • Relationship of unit triangle to building • Sample rafter calculation

21. Hip Rafter Theory and Sample Calculations

Common rafter run vs. hip rafter run • Common rafter slope vs. hip rafter slope • Ridge shortening and cheek cut necessity • Rafter dropping or backing for sheathing • Hip projection and overhang determination • Calculation of hip rafter length

22. Hip Ridge Theory and Lengthening

Styles of hip-ridge intersections • Determining theoretical hip ridge length • Additions to length based on style of construction

23. Valley Jack Theory and Layout

Jacks as common rafters • Determining the run of jacks and calculating length • Common difference in length • Cheek cuts

24. Common Rafter Layout Using the Calculator Method

Gathering information and starting measurements • Ridge cut line and ridge shortening • Sample length calculation • Bird's mouth plumb and level lines • Tails plumb and level lines

25. Common Rafter Assembly

Temporary work surface • Layout ridge with desired gable end overhang • Assemble ridge and two rafters • Raise into position and installing opposing rafters • Bracing this five member assembly plumb • Filling in rafters and continuing for the length of the building

26. Hip Rafter Layout

Ridge plumb line • Ridge shortening • Cheek cuts • Length determination • Bird's mouth plumb line • Seat cut level line • Dropping rafter at seat cut • Tail plumb line • Soffit cut line

27. Hip Jack Theory and Layout

Jacks as common rafters • Determining the run of jacks and calculating length • Common difference in length • Bird's mouth and tail cuts same as common • Cheek cuts

28. Valley Rafter Theory and Layout

Valley similarities to hip • Layout of ridge plumb lines • Length determination • Bird's mouth plumb and level lines • Seat plumb line adjustment • Typical omission of tail

29. Hanging a Prehung Door

Setup for installation • Leveling head jamb • Plumbing and centering side jambs • Shimming at lockset and hinges • Checking door operation

30. Starting 3-tab Shingles

Installing underlayment and metal drip edge • Cutting eaves starter strips • Cutting rake starter pieces • Beginning shingles at rake edge • Beginning shingles at the center of the roof

31. Scribing Concepts

Setup of panel plumb against an out-of-plumb corner • Adjusting scriber • Scribing with level scriber • Scribing a board to existing siding

32. Installing Door Casing

Marking reveal • Cutting head casing • Cutting side casing

33. Installing Baseboard

Choosing corner to begin • First piece cut square both ends • Second piece cut square and coped • Piecing on a long wall • Last piece

34. Installing Window Trim

Window stool • Apron and its return • Head casing • Side casing

35. Installing Wood Strip Flooring

Snapping line and first two courses • Racking floor strips with little waste • Installation • Completing last course

36. Stair Rise and Run

Component parts and names • Styles of stair construction • Determining unit rise and unit run • Determining appropriate stair angle • Styles of handrails

37. Stairwell Length and Headroom

Checking for proper headroom • Determining length of stairwell • Stringers and carriages

38. Stair Sample Calculation

Example from given information

39. Installing Kitchen Cabinets

Layout lines • Installing upper cabinet • Scribing base cabinet to floor and wall • Setting countertop

Appendix B Video Titles

Appendix B lists corresponding numbered video titles and sections as applicable

Video #1: The Building Layout Process

Section 1: The Building Permit
Section 2: Preparing the Lot and Laying Out the Site
Section 3: Establishing the Location of a House Using Front, Rear, and Side Set Backs
Section 4: Using a Transit Level
Section 5: 3—4—5 Method to Locate a Corner
Section 6: General Information on Installing Batter Boards
Section 7: Leveling a Ledger Using a Water Level
Section 8: Leveling a Ledger Using a Laser Level
Section 9: Stretching the Layout Lines
Section 10: Wrapping It Up

Video #2: Setting the Forms

Section 1: Setting the Forms
Section 2: Marking Out the Intersecting Walls
Section 3: Locating Placement of Piping, Digging Trenches, and Footers
Section 4: Termites/Placing Vapor Barrier
Section 5: Placing Reinforcing Steel
Section 6: Marking the Anchor Bolt Locations
Section 7: Placing Welded Wire Fabric and Glass Fibered Concrete
Section 8: Estimating Concrete
Section 9: Last Minute Preparations for Concrete
Section 10: Safety Guidelines for Working With Concrete
Section 11: Placement of Concrete
Section 12: Wrapping It Up

Video #3: Wood Floor Systems/Wood Frame Walls

Section 1: Installing the Sill Plate
Section 2: Marking Locations for Floor Joists, Floor Openings, and Partition Supports
Section 3: Installing Girders
Section 4: Installing Floor Joists
Section 5: Applying the Subflooring
Section 6: Wood in Construction
Section 7: Overview of Wall Components
Section 8: Pre-Assembling the Wall Components
Section 9: Marking the Placement of the Walls
Section 10: Assembling the Wall Components
Section 11: Placing Soleplates and Top Plates
Section 12: Rough Openings
Section 13: Raising the Walls
Section 14: Wrapping It Up

Video #4: Truss Installation

Section 1: Components of a Common Truss
Section 2: Laying Out Trusses
Section 3: Using Step and Portable Ladders
Section 4: Fastening Trusses Together
Section 5: The Role of the Crane in Truss Installation
Section 6: Preparing to Lift Trusses with a Crane
Section 7: Bracing the Trusses
Section 8: Plumbing, Spacing, and Anchoring Trusses
Section 9: Subfacia
Section 10: Final Steps in Truss Installation
Section 11: Wrapping It Up

Video #5: Window & Door Installation

Section 1: Window Components
Section 2: Window Opening
Section 3: Installing a Window Unit
Section 4: Use of Wood Bucks in Concrete Construction
Section 5: Door Components
Section 6: Types of Doors
Section 7: Installing a Door
Section 8: Wrapping It Up

Video #6: Interior Partition Framing

Section 1: Layout and Prep Work
Section 2: Introduction
Section 3: Non-Bearing Partitions
Section 4: Laying and Placing Studs for Interior Partitions
Section 5: Placing Trimmers, Headers, Cripples, and Partition Intersections
Section 6: Raising the Partition
Section 7: Metal Studs
Section 8: Wrapping It Up

Video #7: Interior Trim

Section 1: Moldings
Section 2: Miter Saw — Safety Tips
Section 3: Miter Saw Components
Section 4: Using a Miter Saw
Section 5: Installing an Interior Door
Section 6: Interior Window Installation
Section 7: Interior Window Casings
Section 8: Apron
Section 9: Installing Base Moldings
Section 10: Wrapping It Up

Video #8: Stairs Construction

Section 1: The Tools Needed for Stair Construction
Section 2: Components of Stairs
Section 3: Determining the Total Rise, Run, and Width of Stairs
Section 4: Laying Out a Cutout Carriage
Section 5: Layout Treads and Risers
Section 6: Determine Thickness of Tread
Section 7: Cutting the Carriage
Section 8: Fasten the Carriages
Section 9: Cutting the Risers and the Tread
Section 10: Installing Risers and Treads
Section 11: Wrapping It All Up

Appendix C Answers to Workbook

Chapter 1—Hand Tools

Chapter Review Questions and Exercises

Completion:

1. A __rabbet__ is an L-shaped cutout along the edge or end of lumber.

2. A __crosscut__ is a cut made across the grain of lumber.

3. One __whets__ a tool by rubbing the tool on a flat sharpening stone.

4. A __plumb bob__ is a pointed weight attached to a line for testing plumb.

5. A __square__ is a tool used to mark a layout and mark angles, particularly 90-degree angles; a term used to describe two lines or sides that meet at a 90-degree angle; and the amount of roof covering that will cover 100 square feet of roof area.

6. The __heel__ is the back end of objects, such as a handsaw or hand plane.

7. The __kerf__ is the width of a cut made with a saw.

8. When something is __level__, it is horizontally aligned perpendicular to the force of gravity.

9. A __dado__ is a cut, partway through and across the grain of lumber.

10. When something is __plumb__, it is vertically aligned with the force of gravity.

11. The __toe__ is the forward end of tools, such as a hand saw and hand plane.

12. A __groove__ is a cut, partway through and running with the grain of lumber.

Selecting the right hammer: job sheet 1:

Common Nail Hammer with Curved Claw—used for general carpentry work and, of course, nail pulling

Rip Hammer with Straight Claw—used for general and heavy carpentry work, framing, ripping

Finishing Hammer—used for cabinet making, finishing, and general carpentry

Ball Peen Hammer—used for riveting, center punching, and bending or shaping soft metal

Hand Drilling Hammer—used for powerful work such as striking masonry nails, steel chisels, and masonry drills

Soft-Face Hammer—for assembling furniture and wood projects, setting dowels, and any task that requires non-marring blows

Tack Hammer—used to drive small nails, furniture upholstering, and more

Brick Hammer—designed for cutting and setting bricks of blocks and for chipping mortar

Drywall (Wallboard) Hammer—used for drywall work, marking wallboard, and making cutouts; it sets nails with dimples for easier finishing; corner nailing (in some versions)

Carpenter's Mallet—for use in furniture assembly, shaping soft sheet metals, or any project that requires non-marring blows

Chapter 2—Power Tools

Chapter Review Questions and Exercises

Completion:

1. A __miter gauge__ is a guide used on the table saw for making miters and square ends.

2. To __rip__ is to saw lumber in the direction of the grain.

3. The __bevel__ is the sloping edge or side of a piece at any angle other than a right angle.

4. The __chamfer__ is an edge or end bevel that does not go all the way across the edge or end.

5. A __crosscut__ is a cut made across the grain of lumber.

6. A __compound miter__ is a bevel cut across the width and also through the thickness of a piece.

7. A __fence__ is a guide for ripping lumber on a table saw.

8. To __miter__ is to cut off the end of a piece at any angle other than a right angle.

9. A __board foot__ is a measure of lumber volume that equals 1 foot square and 1 inch thick or any equivalent lumber volume; it is sometimes represented by the letter M.

10. __Medullary rays__ are bands of cells radiating from the cambium layer to the pith of a tree to transport nourishment toward the center.

11. __Millwork__ is any wood products that have been manufactured, such as moldings, doors, windows, and stairs for use in building construction; sometimes called joinery.

12. __Plain-sawing__ is a method of sawing lumber that produces flat grain where annular rings tend to be parallel to the width of the board.

13. __On center (OC)__ is the distance from the center of one structural member to the center of the next one.

14. Trees that are cone-bearing are called __coniferous__; also known as evergreen trees.

15. __Sapwood__ is the outer part of a tree just beneath the bark containing active cells.

16. A __sawyer__ is a person whose job is to cut logs into lumber.

17. __Tempering__ is a treatment process in which the material is made harder and stronger.

18. __Timbers__ are large pieces of lumber over 5 inches in thickness and width.

19. Trees that shed leaves each year are called __deciduous__.

20. __Finger jointing__ is the process in which shorter lengths are glued together using deep, thin V grooves, resulting in longer lengths.

21. __Headers__ are framing members placed at right angles to joists, studs, and rafters to form and support openings.

22. A __panel__ is a large sheet of building material that usually measures 4 × 8 feet.

23. __Quarter-sawing__ is a method of sawing lumber that produces a close grain pattern in which the annular rings tend to be perpendicular to the width of the board.

24. The __pith__ is the small, soft core at the center of a tree.

Chapter 3—Wood and Wood Products

Chapter Review Questions and Exercises

Completion:

1. __Air drying__ is a technique for removing water from lumber using natural wind currents.

2. __Annular rings__ are the rings seen when viewing a cross-section of a tree trunk; each ring constitutes one year of tree growth.

3. The __cambium layer__ is a layer just inside the bark of a tree where new cells are formed.

4. __Heartwood__ is the wood in the inner part of a tree, usually darker and containing inactive cells.

5. __Dimension__ is a term used to define a measurement of an item; also used to refer to all 2× lumber used in framing.

6. __Lumber__ is a general term for wood that is cut from a log to form boards, planks, and timbers.

7. A __dry kiln__ is a large oven used to remove water from lumber.

8. __Board__ is lumber usually less than 2 inches thick.

Calculating board feet: job sheet 1:

Number of Pieces	Length	Width	Thickness	Number of Board Feet
23	8	6	1	92
12	10	8	2	160
45	8	4	2	240
5	12	10	1	50
120	8	2	2	320
37	10	10	2	616 2/3
56	12	12	1	672
12	8	6	2	96
25	8	4	1	66 2/3

Chapter 4—Fasteners

Chapter Review Questions and Exercises

Completion:

1. An __anchor__ is a device used to fasten structural members in place.
2. __Face nailing__ is the method of driving a nail straight through a surface material into a supporting member.
3. __Mastic__ is a thick adhesive.
4. A __brad__ is a thin, short, finishing nail.
5. A __box nail__ is a thin nail with a head, usually coated with a material to increase its holding power.
6. A __duplex nail__ is a double-headed nail used for temporary fastening such as in the construction of wood scaffolds.
7. A __galvanized__ fastener is protected from rusting by a coating of zinc.
8. __Electrolysis__ is the decomposition of one of two unlike metals in contact with each other in the presence of water.
9. A __finish nail__ is a thin nail with a small head designed for setting below the surface of finish material.
10. __Penny (d)__ is a term used in designating nail sizes.
11. __Toenailing__ is a method of driving a nail diagonally through a surface material into a supporting member.

Selecting the correct nail: job sheet 1:

Common Nail—the most often-used of all nails, useful for most applications where the special features of the other nail types are not needed

Box Nails—used for boxes and crates

Finishing Nails—can be driven below the surface of the wood and concealed with putty so that it is completely hidden

Casing Nails—used for installing exterior doors and windows

Duplex Nails—used for temporary structures, such as locally built scaffolds

Roofing Nails—used for installing asphalt and fiberglass roofing shingles

Masonry Nails—used when nailing into concrete or masonry

Selecting the correct screw: job sheet 2:

Drywall Screws—used to attach drywall to wall studs

Sheet metal Screws—used to fasten metal to wood, metal to metal, plastic, or other materials. Sheet metal screws are threaded completely from the point to the head, and the threads are sharper than those of wood screws. Machine screws are for joining metal parts, such as hinges to metal door jambs.

Particleboard and Deck Screws—corrosion-resistant screws used for installing deck materials and particleboard

Lag Screws—used for heavy holding, driven in with a wrench rather than a screwdriver

Chapter 5—Blueprints, Codes, and Building Layout

Chapter Review Questions and Exercises

Completion:

1. A __plan__ is an architectural drawing or an object drawn as viewed from above.

2. A __detail__ is a close-up view of a plan or section.

3. The __foundation__ is that part of a wall on which the major portion of the structure is erected.

4. A __ledger__ is a temporary or permanent supporting member for joists or other members running at right angles; horizontal member of a set of batter boards.

5. The __elevation__ is a drawing in which the height of the structure or object is shown; also, the height of a specific point in relation to another reference point.

6. A __laser__ is a concentrated, narrow beam of light; optical leveling and plumbing instrument used in building construction.

7. A __section__ is a drawing showing a vertical cut-view through an object or part of an object.

8. The __Pythagorean theorem__ is a mathematical expression that states the sum of the square of the two sides of a right triangle equals the square of the diagonal side.

Chapter 6—Concrete Form Construction

Chapter Review Questions and Exercises

Completion:

1. __Reinforcing rods__, also called rebar, are steel bars placed in concrete to increase tensile strength.

2. __Portland cement__ is a fine gray powder that when mixed with water forms a paste that sets rock hard; an ingredient in concrete.

3. A __stud__ is a vertical framing member in a wall running between plates.

4. The __rise__ is the vertical distance of the flight; in roofs, the vertical distance from plate to ridge; may also be the vertical distance through which anything rises.

5. The __run__ is the horizontal distance over which rafters, stairs, and other like members travel.

6. __Concrete block__ is a concrete masonry unit (CMU) used to make building foundations, typically measuring 8″ × 8″ × 16″.

7. __Mortar__ is a mixture of Portland cement, lime, sand, and water used to bond masonry units together.

8. A __gusset__ is a block of wood or metal used over a joint to stiffen and strengthen it.

9. __Concrete__ is a building material made from portland cement, aggregates, and water.

10. A __pilaster__ is a column built within and usually projecting from a wall to reinforce the wall.

11. A __girder__ is a heavy timber or beam used to support vertical loads.

12. A __vapor retarder__, also called a vapor barrier, is a material used to prevent the passage of moisture.

13. A __buck__ is a rough frame used to form openings in poured concrete walls.

14. The __frost line__ is the depth to which the ground typically freezes in a particular area; footings must be placed below this depth.

15. A __scab__ is a length of lumber or material applied over a joint to stiffen and strengthen it.

16. A __spreader__ is a strip of wood used to keep other pieces a desired distance apart.

Estimating the amount of concrete for a slab: job sheet 1:

Breaking the slab into two areas produces the following results:

Area 1:
35′ − 10″ × 15′ − 7 1/2″ = Area
35.833333 feet × 15.625 feet = Area
564.82 SF = Area 1

Area 2:
20′ − 5″ × 15′ − 6″ = Area
20.4166 feet × 15.5 feet = Area
316.4573 SF = Area 2

Adding the two areas together

Area^1 + Area^2 = Area Total
564.82 + 316.4573 = Area Total
881.2773 SF = Area Total

Calculating the volume

881.2773 SF * .5 ft = CF (volume)
440.6386 cubic feet of concrete

Converting to cubic yards

440.6386/27 = cubic yards
16.339 = cubic yards

Estimating the amount of concrete for a footer: job sheet 2:

The second to last division is problematic. Divide the subtotal 117,436 cubic inches by 1,728 cubic inches/cubic foot to yield 67.9606 cubic feet. In the existing calculation, 117,436 cubic inches is divided by 144 square inches to yield 815.5278. One more division by 12 is needed to yield the desired 67.906 cubic feet. One then divides 67.906 cubic feet by 27 cubic feet/cubic yard to yield the final answer of 2.517 cubic yards.

Feet	Inches	Total Inches
15	8	188
10	4	124
25	2	302
20	9	249
40	10	490
31	2	374
	subtotal footing length	1727
	footing	124344
	keyway	6908
	subtotal	117436
	cubic feet of concrete	67.960
	cubic yards of concrete	2.517

Chapter 7—Floor Framing

Chapter Review Questions and Exercises

Completion:

1. A __balloon frame__ is a type of frame in which studs are continuous from foundation sill plate to roof.
2. A __band joist__ is the member used to stiffen the ends of floor joists where they rest on the sill.
3. __Draftstops__, also called firestops, are material used to reduce the size of framing cavities in order to slow the spread of fire; in a wood frame, they consist of full-width dimension lumber blocking between studs.
4. __Dimension lumber__ is a term used to describe wood that is sold for framing and general construction.
5. A __plate__ is a top or bottom horizontal member of a wall frame.
6. A __column__ is a large vertical member used to support a beam or girder.
7. A __header__ is a member placed at right angles to joists, studs, and rafters to form and support openings in a wood frame.
8. __Bridging__ is diagonal braces or solid wood blocks between floor joists used to distribute the load imposed on the floor.
9. An __anchor bolt__ is a long metal fastener with a threaded end used to secure materials to concrete.
10. __Pressure treating__ is a treatment given to lumber that applies a wood preservative under pressure.
11. __Flush__ is a term used to describe when surfaces or edges are aligned with each other.
12. A __joist__ is a horizontal framing member used in a spaced pattern that provide support for the floor or ceiling system.
13. __Girders__ are heavy beams that support the inner ends of floor joists.
14. __Linear feet__ a measurement of length.
15. __Masonry__ is any construction of stone, brick, tile, concrete, plaster, and similar materials.
16. __Platform framing__ is a method of wood frame construction in which walls are erected on a previously constructed floor deck or platform.
17. A __post__ is a vertical member used to support a beam or girder.
18. __Sheathing__ is boards or sheet material that is fastened to joists, rafters, and studs and on which the finish material is applied.
19. A __ribbon__ is a narrow board let into studs of a balloon frame to support floor joists.

20. Subfloor material is used as the first floor layer on top of joists.

21. Sill sealer is material placed between the foundation and the sill to prevent air leakage.

22. Tail joists are shortened on center joists running from a header to a sill or girder.

23. Shims are thin, wedge-shaped pieces of material used behind pieces for the purpose of straightening them or for bringing their surfaces flush.

24. A termite shield is a metal flashing plate over the foundation to protect wood members from termites.

25. The sill is the first horizontal wood member resting on the foundation supporting the framework of a building; also, the lowest horizontal member in a window or door frame.

26. A trimmer is a joist or stud placed at the sides of an opening running parallel to the main framing members.

Determining the joist length of a cantilevered joist: job sheet 1:

Over All Dimension	Extended Over Foundation
12	4
18	6
9	3
16.5	5.5
30	10

Chapter 8—Wall and Ceiling Framing

Chapter Review Questions and Exercises

Completion:

1. A gable end is the triangular-shaped section on the end of a building formed by the common rafters and the top plate line.

2. Gypsum board is a sheet product made by encasing gypsum in a heavy paper wrapping used to create the wall surface.

3. Blocking are pieces of dimension lumber installed between joist and studs for the purposes of providing a nailing surface for intersecting framing members.

4. Loadbearing is a term used to describe a structural member that carries weight from another part of the building.

5. A soffit is the underside trim member of a cornice or any such overhanging assembly.

6. Backing strips are blocks of wood installed in walls or ceilings for the purpose of fastening or supporting trim or fixtures.

7. A joist hanger is a metal stirrup used to support the ends of joists that do not rest on top of a support member.

Types of solid and built up headers: job sheet 1:

See Figure 8-3 in Residential Construction Carpentry.

Window framing: job sheet 2:

See Figure 8-4 in Residential Construction Carpentry.

Door framing: job sheet 3:

See Figure 8-5 in Residential Construction Carpentry.

Chapter 9—Scaffolding, Ladders, and Sawhorses

Chapter Review Questions and Exercises

Multiple choice:

Identify the choice that best completes the statement or answers the question.

1. The base of a ladder should be 1 foot away from the vertical surface against which it is placed for every __________ feet in height.
   a. 2
   b. 3
   c. 4
   d. 5

2. Employees must be able to safely access any level of scaffold that is __________ feet above or below an access point.
   a. 2
   b. 3
   c. 4
   d. 5

3. Fall protection must be provided for workers on scaffolds when they are __________ feet or more above a lower level.
   a. 6
   b. 10
   c. 12
   d. 15

4. When using ladders near electrical equipment, ladders made of __________ should be used.
   a. wood
   b. aluminum
   c. fiberglass
   d. either A or C

5. The most common type of scaffolding used in the construction industry is the __________.
   a. pump jack
   b. fabricated frame
   c. moveable
   d. adjustable

Short answer:

1. Explain what a yellow tag on a scaffold means.
   A scaffold does not meet all safety requirements, but may be used with a fall protection harness.

2. Explain what a red tag means when it is attached to a scaffolding.
   The scaffold is not safe to use.

Completion:

1. A competent person is a designated person on a job site who is capable of identifying hazardous or dangerous situations and has the authority to take prompt corrective measures to eliminate them.

2. An erector is a worker whose responsibilities include safe assembly of scaffolding.

3. A cleat is a small strip of wood applied to support a shelf or similar piece.

4. Users are people who work on scaffolding.

5. A crib is a heavy wood block and framing used as a foundation for scaffolding.

Chapter 10—Roof Framing

Chapter Review Questions and Exercises

Completion:

1. A gambrel roof is a type of roof that has two slopes of different pitches on each side of center.

2. A cheek cut is a compound miter cut on the end of certain roof rafters.

3. A hip jack is a rafter running between a hip rafter and the wall plate.

4. A dormer is a structure that projects out from a sloping roof to form another roofed area to provide a surface for the installation of windows.

5. A gable roof is a common type of roof that pitches in two directions.

6. A fascia is a vertical member of the cornice finish installed on the bottom end of rafters.

7. A hip rafter extends diagonally from the corner of the plate to the ridge at the intersection of two surfaces of a roof.

8. A valley cripple jack rafter is a rafter running between two valley rafters.

9. A valley jack rafter is a rafter running between a valley rafter and the ridge.

10. Lateral is a direction to the side at about 90 degrees.

11. A lookout is horizontal framing pieces in a cornice, installed to provide fastening for the soffit.

12. An intersecting roof is the roof of irregular-shaped buildings; valleys are formed at the intersection of the roofs.

13. A valley rafter is the rafter placed at the intersection of two roof slopes in interior corners.

14. The rake is the sloping portion of the gable ends of a building.

15. The valley is the intersection of two roof slopes at interior corners.

16. A tail cut is a cut on the extreme lower end of a rafter.

17. A shed roof is a type of roof that slopes in one direction only.

18. A mansard roof is a type of roof that has two different pitches on all sides of the building, with the lower slopes steeper than the upper.

19. A hip-valley cripple jack rafter is a short rafter running parallel to common rafters, cut between hip and valley rafters.

Truss component identification: job sheet 1:

See Figure 10-42, in residential construction carpentry.

Truss component identification: job sheet 2

See Figure 10-41, in residential construction carpentry.

Chapter 11—Windows and Doors

Chapter Review Questions and Exercises

Completion:

1. A hopper window is a type of window in which the sash is hinged at the bottom and swings inward.
2. An astragal is a semicircular molding often used to cover a joint between two doors.
3. Low emissivity glass (low E) is a coating on double-glazed windows designed to raise the insulating value by reflecting heat back into the room.
4. Molding is a decorative strip of wood used for finishing purposes.
5. A muntin is a slender strip of wood between lights of glass in windows or doors.
6. A rail is the horizontal member of a frame.
7. A sash is that part of a window into which the glass is set.
8. A back miter is an angle cut starting from the end and going back on the face of the stock.
9. An escutcheon is a protective plate covering the knob or key hole in doors.
10. A bay window is a window, usually three-sided, that projects out from the wall line.
11. Casing is molding used to trim around doors, windows, and other openings.
12. Double-acting doors/hinged are doors that swing in both directions, or the hinges used on these doors.
13. Glazing is the act of installing glass in a frame.
14. A mullion is a vertical division between window units or panels in a door.
15. A light is a pane of glass or an opening for a pane of glass.
16. Housewrap is a type of building paper with which the entire exterior sidewalls of a building are covered.
17. Extension jambs are strips of wood added to window jambs to bring the jamb edge flush with the wall surface in preparation for casing.
18. Insulated glass is multiple panes of glass fused together with an air space between them.
19. A stile is the outside vertical members of a frame, such as in a paneled door.
20. Weather stripping is a narrow strip of material applied to windows and doors to prevent the infiltration of air and moisture.
21. Wind is a defect in lumber caused by a twist in the stock from one end to the other; also, a twist in anything that should be flat.

Short answer:

1. How are items of finish carpentry and millwork generally measured?

   Items of finish carpentry and millwork generally are classified and measured separately according to materials, size, and method of installation, i.e. wall-mounted items would be measured separately from floor mounted or ceiling mounted.

2. Describe how doors and doorframes are measured.

   Doors and doorframes are enumerated stating their size and type of material. Windows are enumerated and fully described including the rough opening sizes of the units.

Hanging an interior door: job sheet 1:

See pages 325–336 Residential Construction Carpentry.

Installing windows: job sheet 2:

See pages 223–224 Residential Construction Carpentry.

Chapter 12—Roofing

Chapter Review Questions and Exercises

Completion:

1. Selvage the unexposed part of roll roofing covered by the course above.

2. Asphalt felt is a building paper saturated with asphalt for waterproofing.

3. An open valley is a roof valley in which the roof covering is kept back from the centerline of the valley.

4. Exposure is the amount that courses of siding or roofing are exposed to the weather.

5. Square is the amount of roof covering that will cover 100 square feet of roof area.

6. An apron is the flashing piece located on the lower side of a roof penetration such as a chimney or dormer.

7. Flashing is the material used at intersections such as roof valleys and dormers and above windows and doors to prevent the entrance of water.

8. A closed valley is a roof valley in which the roof covering meets in the center of the valley, completely covering the valley.

9. A drip edge is a metal edging strip placed on roof edges to provide a support for the overhang of the roofing material.

10. Electrolysis accelerated oxidation of one metal because of contact with another metal in the presence of water.

Estimating shingles: job sheet 1:

26.15 squares

Installing asphalt shingles: job sheet 2

See pages 352–355 Residential Construction Carpentry.

Chapter 13—Siding and Decks

Chapter Review Questions and Exercises

Completion:

1. The drip is the part of an exterior finish that projects below another to cause water to drop off instead of running back against and down the wall.

2. A downspout is a vertical member used to carry water from the gutter downward to the ground; also called a leader.

3. A soffit is the horizontal, underside trim member of a cornice or other overhanging assembly.

4. Striated is the term to describe a finish material with random and finely spaced grooves running with the grain.

5. Blind nailing is a method of fastening that conceals the fastener.

6. A story pole is a narrow strip of wood used to lay out the insulation heights of material such as siding or vertical members of a wall frame.

7. Battens are thin, narrow strips typically used to cover joints in vertical boards.

8. Corner boards are finish trim members used at the intersection of exterior walls.

9. A baluster is the vertical member of a stair rail, usually decorative and spaced closely together.

10. A plancier is the finish member on the underside of a box cornice, also called a soffit.

11. Cornice is a general term used to describe the part of the exterior finish where the walls meet the roof.

12. Eaves are the lower part of the roof that extend beyond the sidewalls.

13. A frieze is that part of the exterior trim applied to cover the joint between the overhanging cornice and the siding.

14. A gutter is a trough attached to an eave used to carry off water.

15. A rake is the portion of the roof that overhangs the gable end.

16. Water tables are exterior trim members that are applied at the intersection of the siding and the foundation that project outward to direct water away from the building.

Installing vinyl siding: job sheet 1:

See pages 394–399 Residential Construction Carpentry.

Installing wood shingles and shakes: job sheet 2:

See pages 392–393 Residential Construction Carpentry.

Chapter 14—Insulation and Wall Finish

Chapter Review Questions and Exercises

Completion:

1. Gypsum board is a panel used as a finished surface material made from a mineral mined from the earth.
2. Condensation is when water, in a vapor form, changes to a liquid due to cooling of the air; the resulting droplets of water that accumulate on the cool surface.
3. Insulation is a material used to restrict the passage of heat or sound.
4. Dew point is the temperature at which moisture begins to condense out of the air.
5. Hardboard is a building product made by compressing wood fibers into sheet form.
6. Wainscoting is a wall finish applied partway up the wall from the floor.
7. An eased edge is an edge of lumber whose sharp corners have been rounded.
8. The R-value is a number given to a material to indicate its resistance to passage of heat.
9. A storm sash is an additional sash placed on the outside of a window to create dead air space to prevent the loss of heat from the interior in cold weather.
10. The face is the best-appearing side of a piece of wood or the side that is exposed when installed.
11. Air infiltration is unwanted movement of air into an insulation layer of conditioned space.
12. A vapor retarder is a material used to prevent the passage of water in the gaseous state.
13. Weather stripping is a narrow strip of thin metal or other material applied to windows and doors to prevent the infiltration of air and moisture.

Using the drywall estimating form: job sheet 2:

11.25	7.5	6.75	5.625
4 × 8	4 × 9	4 × 10	4 × 12

Chapter 15—Interior Finish

Chapter Review Questions and Exercises

Completion:

1. An apron is a piece of the window trim used under the stool.
2. A stool is the bottom horizontal member of interior window trim that serves as the finished window sill.
3. A spline is a thin, flat strip of wood inserted into the grooved edges of adjoining pieces.
4. A compound miter is a bevel cut across the width and also through the thickness of a piece.

Calculating ceiling tiles: job sheet 1:

Room			Ceiling		
Length (ft)	Width (ft)	Area (SF)	Tile Dimensions	Area (SF)	Number of Tiles Needed
8′-3″ ⇒ 9	8′-9″ ⇒ 9	81	2 × 4	8	10.1 ⇒ 11
9′-9″ ⇒ 10	10′-4″ ⇒ 11	110	2 × 2	4	27.5 ⇒ 28
13′-9″ ⇒ 14	9′-4′ ⇒ 10	140	2 × 2	4	35
11′-5″ ⇒ 12	11′-5″ ⇒ 12	144	2 × 2	4	36
12′-3″ ⇒ 13	12′-4″ ⇒ 13	169	2 × 4	8	21.1 ⇒ 22
16′-9″ ⇒ 17	16′-6″ ⇒ 17	289	2 × 4	8	36.1 ⇒ 37

Chapter 16—Stair Framing and Finish

Chapter Review Questions and Exercises

Completion:

1. The __handrail__ is a railing on a stair intended to be grasped by the hand to serve as a support and guard.

2. The __rake__ is the sloping portion of trim, such as on grade ends of a building or stair.

3. A __baluster__ is a vertical member of a stair rail, usually decorative and spaced closely together.

4. The __balustrade__ is the entire stair rail assembly, including handrail, balusters, and posts.

5. A __newel post__ is an upright post supporting the handrail in a flight of stairs.

6. A __stairwell__ is an opening in the floor for climbing or descending stairs or the space of a structure where the stairs are located.

7. A __shank hole__ is a hole drilled for the thicker portion of a wood screw.

Determine the rise and run of a set of stairs: job sheet 2:

Total Rise = 10 feet
Number of Risers = 15.48 or 15
Number of Treads = 14

Stair terminology: job sheet 3:

Stringer or Stair Jack—support for the treads

Kick Block or Kicker—used to keep bottom of stringer from sliding on the floor

Headroom—vertical distance measured from tread nosing to wall or floor above stairs

Handrail—railing on which to slide hand

Guardrail—railing between stair opening

Chapter 17—Cabinets and Countertops

Chapter Review Questions and Exercises

1. A __j-roller__ is a 3-inch-wide rubber roller used to apply pressure over the surface of contact cement-bonded plastic laminates.

2. __Postforming__ is the method used to bend plastic laminate to small radii.

3. A __pivot__ is a point of rotation.

4. __Face frame__ is a framework of narrow pieces on the front of a cabinet making the door and drawer opening.

5. A __pilot__ is a guide on the end of edge-forming route bits used to control the amount of cut.

6. A __gain__ is a cutout made in a piece to receive another piece, such as a cutout for a butt hinge.